P9-CPX-118

Ways to Be a Better SALTWATER FISHKEEPER

Dave Garratt, Tim Hayes, Tristan Lougher and Dick Mills

FIREFLY BOOKS

A Firefly Book

Published by Firefly Books Ltd. 2005

First printing

Publisher Cataloging-in-Publication Data (U.S.)

Garratt, Dave.

500 ways to be a better saltwater fishkeeper/ Dave Garratt ... [et al.].

[128] p. : col. photos. ; cm.

Summary: Reference that features 500 practical tips for saltwater fishkeepers. Includes information on aquariums, heating and lighting, water management, fish, invertebrates, feeding, ongoing care, breeding and healthcare.

ISBN 1-55407-047-3

1. Marine aquariums. 2. Marine aquarium fishes. I. Title.

639.34/2 dc22 SF457.1.G377 2005

Library and Archives Canada Cataloguing in Publication

500 ways to be a better saltwater fishkeeper/ Dave Garratt ... [et al.]

ISBN 1-55407-047-3

1. Marine aquariums. 2. Marine aquarium fishes. I. Garratt, Dave II. Title: Five hundred ways to be a better saltwater fishkeeper.

SF457.1.F59 2005 639.34'2 C2005-902516-6

Published in the United States by
Firefly Books (U.S.) Inc.
P.O. Box 1338, Ellicott Station
Buffalo, New York 14205

Published in Canada by
Firefly Books Ltd.
66 Leek Crescent
Richmond Hill, Ontario L4B 1H1

Created and compiled: Ideas into Print, Claydon, Suffolk IP6 0AB, England

Design and prepress: Stuart Watkinson, Ayelands, New Ash Green, Kent DA3 8JW, England

Editorial consultant: Tim Hayes

Production management: Consortium, Poslingford, Suffolk CO10 8RA, England

Print production: SNP Leefung

Printed in China

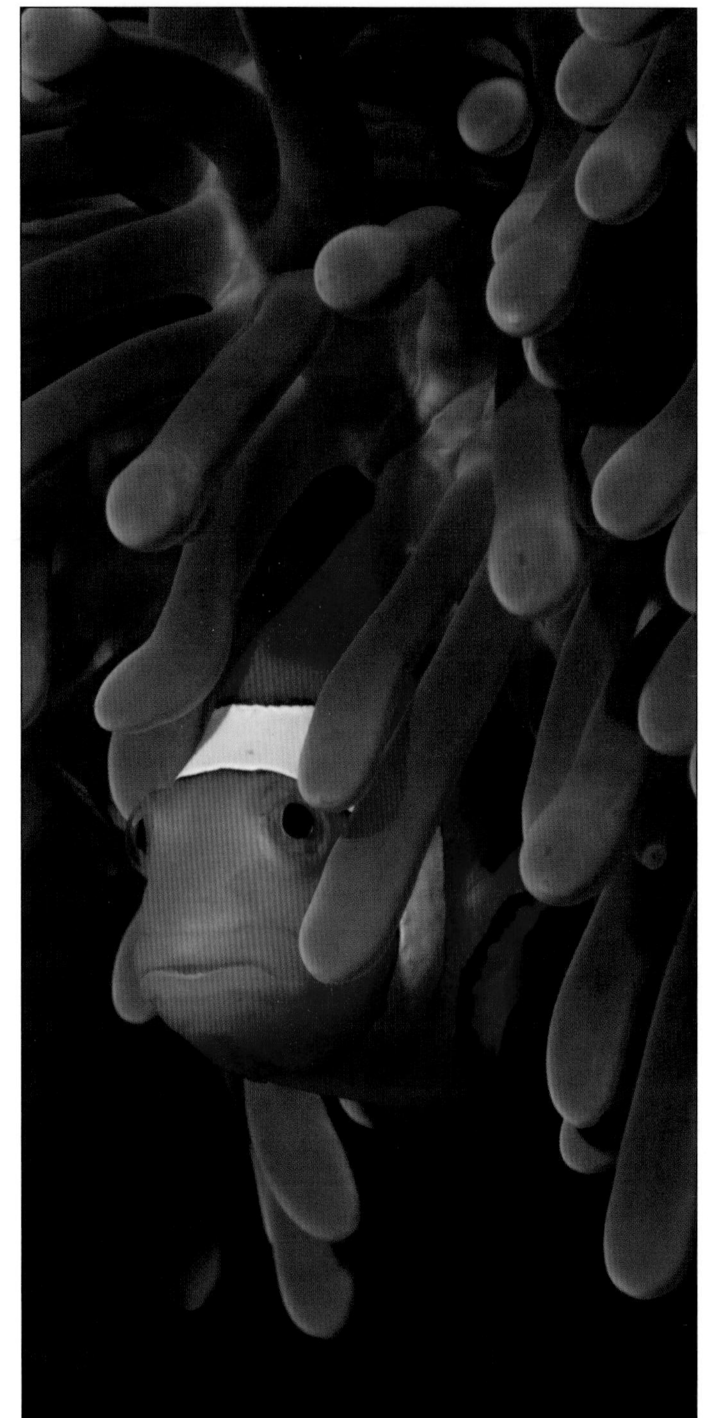

Contents

1 Are saltwater fish more difficult to keep than freshwater fish?

Saltwater, or marine, fish are generally considered to be more difficult to keep than freshwater fish for a couple of very good reasons. One is the stability of the coral reef environment. Such stability means that there have been no pressures on saltwater fish to evolve mechanisms to deal with any variations, and therefore they do not respond well to unstable conditions in captivity. Secondly, pathogenic organisms do not abound in great numbers on the reef so, once again, saltwater fish are ill-equipped to deal with a disease outbreak in the close confines of the aquarium.

Tip 1 *A reef aquarium re-creates a highly stable environment.*

2 Learn as much as you can before starting out with saltwater fishkeeping

Make sure you have a good understanding of what is happening in your aquarium. Find out all you can about key issues, including maturation of a biological filter, fish and invertebrate compatibility, correct diet and feeding, the natural behavior of the species you hope to keep and stocking levels. The knowledge you acquire will pay great dividends in the future.

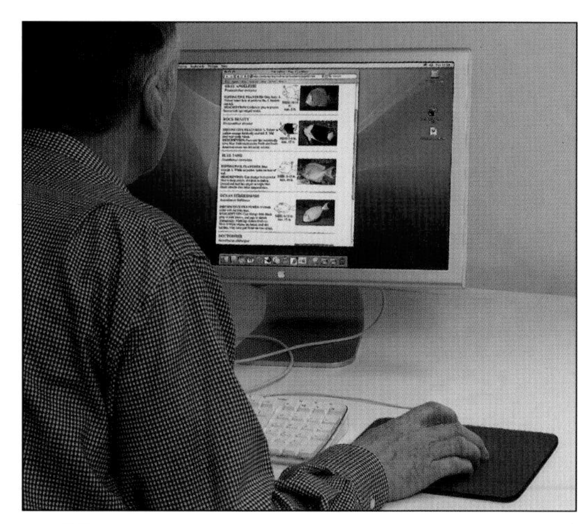

Tip 3 *The internet is a great source of information.*

3 Use all the research avenues open to you, including the internet

One of the keys to success when contemplating entering the saltwater aquarium hobby is research. Learn as much as you can about the animals you are considering keeping. Put together a library of good, up-to-date books, read magazines and use the resources available on the internet. However, when taking advice from the internet be wary of the source; not everybody online is a bona fide expert. And do not allow yourself to be sidetracked by information of a more speculative nature aimed at the advanced aquarist.

4 In what circumstances should I set up a fish-only aquarium?

A fish-only tank will usually contain either larger species or fish that would prey on invertebrate animals such as corals, crustaceans, molluscs, etc. Often, aquarists setting up a fish-only aquarium will use inert rock for aquascaping due to the perceived likelihood that at some point copper-based medications will be used. These treatments are toxic to invertebrates. (See also Tip 491.)

Tip 7 Dardanus guttatus, *a hermit crab compatible with large fish.*

5 Can I put live rock into a fish-only aquarium?

In one variation on the fish-only tank, live rock is used instead of inert rockwork. This precludes the use of copper-based medications because of the large populations of small invertebrate animals inhabiting the rock. However, as well as playing an important role in filtration, live rock can provide a healthier environment for fish, because they can browse on the invertebrate life in the rock. By providing a more favorable environment, you reduce the risk of having to use invertebrate-toxic medication.

6 A fish-only aquarium does not always mean fish only!

Depending on the fish species being kept, it may be possible to include some invertebrates in a nominally fish-only aquarium. Some fish will eat corals but not crustaceans or molluscs; some may eat crustaceans or molluscs but not corals. So, with some careful research, you may be able to identify invertebrates of one species or another that can be safely introduced into a "fish only" system.

7 Invertebrates may be safe with fish, but are the fish safe from them?

Some species of invertebrates may be tough enough, or big enough, to be safe with certain, otherwise predatory, fish. Large brittle stars (such as the green brittle star, *Ophiarachna incrassata*), large hermit crabs (such as *Dardanus* spp.) and carpet anemones (*Heteractis* spp.) are all good examples. But larger invertebrates may be a threat to certain species of fish. Always research compatibility before you add new animals to your aquarium. (See also Tips 313–336.)

8 Protect invertebrates when treating fish with medication

If you do maintain some invertebrates in a predominantly fish-themed aquarium, you can always remove them to a holding tank if you need to treat the fish with an invertebrate-toxic medication. After treatment with a copper-based medication, test for any residual levels. Even if there is no trace, it is a good idea to use one of the commercially available media designed to remove copper, and then test again before reintroducing the invertebrates.

9 What is the definition of a reef aquarium?

At its most basic, a reef aquarium is one where the emphasis is on keeping a display of corals with a few small, compatible fish, plus sundry, nonpredatory, mobile invertebrates. In terms of lighting and water flow, the emphasis will be on providing the best conditions for maintaining the corals.

10 But what is a reef? Is it the same all over the world?

There is no such thing as a single generic "reef." Scientifically speaking, there are about five different types of coral reef, which can be further broken down into at least 10 different major reef zones. Furthermore, there are a few non-reef zones, such as seagrass beds and mangrove swamps. Each of these zones enjoys different conditions in terms of the amount of light it receives, the depth and movement of the water, what species of fish and invertebrates are present, whether it is rocky or sandy, etc.

11 So what do I put in my reef aquarium?

The animals found in the saltwater hobby are collected from all the different reef zones. One of the first things you should do is research the zone your animals come from and then try to replicate that particular environment in your reef tank. At the simplest level, it makes good sense to keep only animals from one of these zones in any individual captive reef as you are then housing a selection of animals with the same requirements in terms of water movement, levels of lighting and substrate.

12 Why not try a geographically themed aquarium?

Not only are the fish and invertebrates offered for sale collected from different reef zones, they are also collected from all over the world. A geographically themed reef would be dedicated to fish and invertebrates collected from, say, the Caribbean, the Red Sea or Hawaii.

Tip 10 *Reefs vary around the world. This is a shallow flat reef on the Great Barrier Reef.*

13 Setting up a biotope aquarium to simulate one environment

This is an aquarium, usually a reef setup, that is well researched, then carefully put together to replicate one particular reef or non-reef zone. The idea here is to re-create the environment as accurately as possible and to keep only the fish and invertebrates that would occur naturally together in your chosen zone.

14 What is the advantage of a saltwater species tank?

Traditionally, a species tank has been used to keep fish that are not compatible with other species. With saltwater species aquariums, the definition can be extended to include a number of differently themed aquariums that limit the number of inhabitants or species. Such a tank might highlight a symbiotic relationship that occurs between two different species or to house small, delicate animals (such as *Periclimenes* species of anemone shrimp and sexy shrimp, *Thor amboinensis*) that might otherwise be eaten by their tankmates. Alternatively, a species tank may house a potentially dangerous animal, such as a predator that presents a risk to other inhabitants, or an animal that, if stressed, has the potential to poison the tank (such as the colorful but deadly sea apple, *Pseudocolochirus* spp.).

Tip 14 *Anemone shrimp* (Periclimenes pedersoni)

15 Does a reef aquarium have to include rocks?

It is not compulsory to have rockwork in your reef tank. By using a deep sand bed and planting seagrasses, macro-algae or mangroves, you can re-create a completely different sort of habitat. A number of corals are termed "free living" and would do better in this environment than in a traditional rock wall reef. If you intend to include mangroves, remember that they will grow out of the water, so plan your lighting accordingly.

Tip 16 *A clownfish with its associated anemone.*

16 Decide which animals you want to keep

One of the first questions to ask yourself is: what are the animals I really want to keep? The answer to this question will often influence how you put your reef aquarium together. For example, if you answer "clownfish and anemones," the reef design will be totally different from one that would suit small-polyped stony corals, such as *Acropora* species. This in turn will dictate your aquascaping, lighting, companion fish and invertebrate species, and water flow requirements.

17 Begin by keeping the less demanding saltwater species

When you start in the saltwater hobby it is easy to be swept away by some of the stunning fish available to you. Why bother with the less striking fish when you can enjoy the majesty of an angelfish or the grace of a butterflyfish? The answer is that these fish are not suitable for the beginner and you should not let your heart rule your head. Do not overlook some of the unsung species, such as the bicolor blenny, *Ecsenius bicolor*. (See also Tip 197.)

Tip 20 *The shortening effect of water reduces the apparent width (front to back measurement) of an aquarium.*

18 Setting up a reef aquarium requires a great deal of patience

If you are starting your first reef aquarium, be patient. You cannot rush the process. Build up the reef over a period of time; it may be 12 months before you can consider it a stable environment. Try to put off adding fish for as long as you can restrain yourself, and concentrate on building up the aquascape. Stick with conservative, easy-to-keep species for the first year while you are learning how to maintain the aquarium.

19 Buy the largest tank possible for a stable marine setup

There are various very attractively shaped tank designs on the market, including bow-fronted tanks. Always choose the largest aquarium you can afford and accommodate. Newcomers should opt for a tank no less than 36 inches (90 cm) long, with a capacity of 36 gallons (136 L). The larger the tank, the less effect any fluctuations in water conditions will have on marine life. Although it is easier to maintain stable water conditions in a large tank and you can keep more fish, bear in mind that the greater the capacity, the more expensive it will be to populate it with live rock and corals if you are setting up a reef aquarium. (See also Tip 259.)

20 Consider the width and depth of a saltwater aquarium

With a saltwater tank, especially a reef tank, you should maximize the width (front to back measurement) to reduce the optical shortening effect that you get in a water-filled aquarium. The depth of the aquarium is important in terms of the lighting. Light reduces in proportion to the square of the distance from its source, so the greater the depth, the more lighting you will need. Also bear in mind that deep tanks can be difficult to service.

Tip 19 *A starter freshwater tank (left) and marine tank compared.*

21 Tank volumes and how to calculate them

Use the following formula to calculate the capacity of your aquarium. The length (L) x the width (W) x the depth (D) in centimeters, divided by 1,000, will give you the volume in liters. To convert to gallons, multiply by 0.22. Live rock and other decor will displace the water; reduce the volume by approximately 10 percent to allow for this, depending on your aquarium setup.

A selection of tank sizes, capacities and weights

Tank size (L x W x D)	Volume of water		Weight of water	
36 x 16 x 15 in. (90 x 40 x 38 cm)	36 gallons	(136 L)	300 lb.	(136 kg)
36 x 18 x 18 in. (90 x 45 x 45 cm)	48 gallons	(182 L)	400 lb.	(182 kg)
39 x 18 x 18 in. (100 x 45 x 45 cm)	54 gallons	(203 L)	450 lb.	(203 kg)
48 x 18 x 18 in. (120 x 45 x 45 cm)	64 gallons	(243 L)	540 lb.	(243 kg)
48 x 24 x 18 in. (120 x 60 x 45 cm)	86 gallons	(324 L)	710 lb.	(324 kg)
60 x 24 x 24 in. (150 x 60 x 60 cm)	143 gallons	(540 L)	1,191 lb.	(540 kg)
72 x 24 x 24 in. (183 x 60 x 60 cm)	175 gallons	(660 L)	1,455 lb.	(660 kg)

22 A fully furnished aquarium is heavy and needs support

Keep the total weight of a display aquarium in mind, especially if the tank is in an upstairs room or in an apartment. Make sure the aquarium is supported firmly, and not only by its stand—its not inconsiderable weight should be distributed evenly over the floor joists. It should also be level in all directions to avoid setting up dangerous stresses in the glass panels, which could then split under water pressure from the inside.

23 Buy another tank for quarantining new stock

While you are choosing your main display aquarium, also buy another smaller tank. A separate quarantine tank is the best insurance you have against inadvertently introducing disease into the aquarium. All new fish stock should be given a minimum two-week period of quarantine before being added to your main collection. (See also Tip 478.)

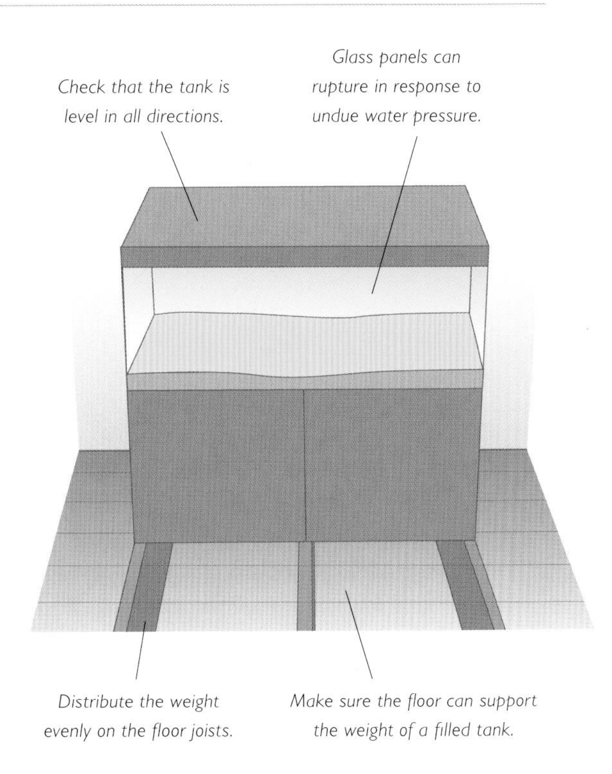

Tip 22 *Tanks are heavy—support them firmly.*

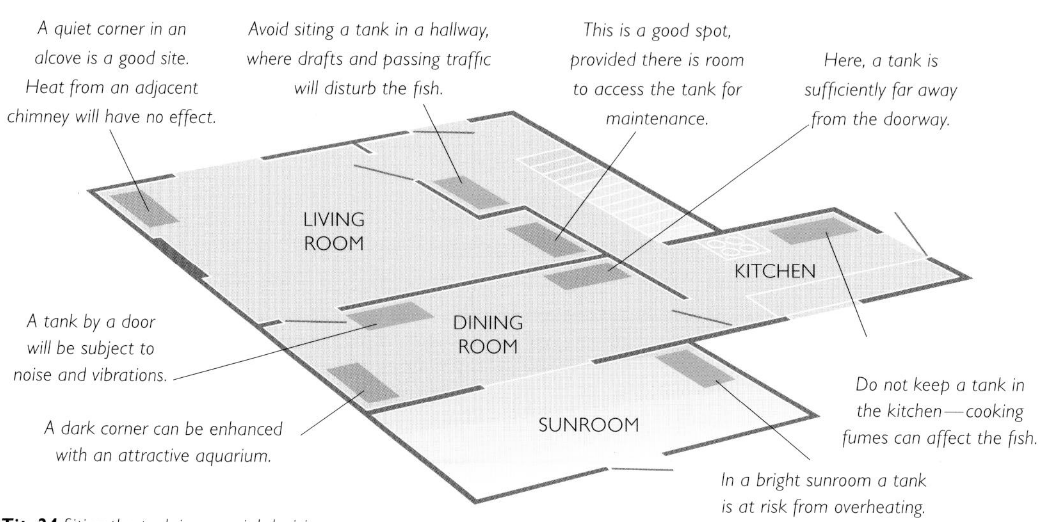

Tip 24 *Siting the tank is a crucial decision.*

24 Finding the best location for a saltwater aquarium

Site the aquarium where it will enjoy stable conditions. Make sure there is easy access to an electricity supply and choose a spot away from drafts and any vibrations from TV or stereo equipment. Do not site the tank in direct sunlight or close to a window in a room that receives the midday sun. Adequate ventilation in the room will help to keep temperatures down in the summer. Avoid a site close to where toxic fumes (e.g., from a central heating duct) could be drawn back into the tank via an air pump. Site the tank away from regular walkways through the house.

25 Allow space for maintenance and keep everything tidy

However neat a job you make of fitting an aquarium into a room, always make sure there is adequate space around it for essential maintenance. With all the ancillary electrical equipment around the aquarium, it makes sense to keep cables not only neat and tidy, but also clearly labeled.

26 Attach the background to the aquarium first

Attaching a suitable background to the external back glass of the aquarium serves two purposes: it can create an illusion of extra depth and also hide what's behind the tank. Be sure to attach the background first before you move the tank into its final position and fill it with water. Try to choose a suitable marine theme (Grecian or Roman columns, sunken galleons and treasure chests notwithstanding) or plain blue or black.

Tip 26 *Attach the background at an early stage.*

Tip 27 *A sump tank with internal dividers.*

27 A sump tank will house the aquarium hardware

To maximize swimming space in the display aquarium, consider housing all the hardware (heaters, filtration equipment, etc.) in a sump beneath the main tank. Whether to take this route is a decision you will need to make at the outset. The only items not likely to be housed in sumps are the small powerheads hidden in the rockwork to provide water flows. (See also Tip 135.)

Tip 29 *A yellow-headed jawfish* (Opistognathus aurifrons) *retreats when disturbed.*

28 Make sure a sump tank system is fail-safe

When running a sump below the display aquarium (or for that matter a refugium higher than the display), it is imperative that the system is properly designed to be fail-safe. In the event of a power failure, the sump must be large enough to accommodate the water that will drain from the display without overflowing. Likewise, when power is restored, siphoning action must restart automatically to prevent the sump running dry and the display from overflowing.

29 Aquarium fish appreciate a quiet life in your home

Provide your fish with a quiet life; it will do wonders for their well-being. Make sure the tank is well protected from inquisitive little hands and teach children not to rush up to the tank or to tap the glass. Do not constantly have your hands in the tank making minor adjustments to the decor.

30 Take nothing for granted when keeping saltwater fish

Always work on the worst possible scenario with saltwater fish. Famous last words include: "that fish will never get through there," "we never have power failures around here" and "I don't need any protection over my pump intakes because my anemone never moves." If it can happen in a saltwater aquarium, it often will. By being a pessimist you can avert disaster.

31 Be prepared to commit time and effort to essential maintenance

With the emphasis on maintaining water quality as the number one consideration, you should fully understand what is involved. There is hardly any margin for error in this respect, so be prepared to devote regular amounts of time to maintenance tasks. (See also Tips 369–391.)

Better Aquarium Furnishings

32 Choosing the right rocks for the saltwater aquarium

For all saltwater aquariums it is better to stick with rock of calcareous origin, such as limestone, rather than risk using unsuitable rock that may react with saltwater and prove to be toxic to livestock.

This ocean rock has been smoothed off.

Calcareous oceanic rock in its natural state is a good foundation for decor in a fish-only aquarium.

***Tip 33** Forms of ocean rock.*

33 Is ocean rock suitable for a saltwater aquarium?

In a fish-only aquarium you can build up the aquascape using ocean rock, which is formed from ancient coral. It is readily available, inexpensive, but dense. In the smaller aquarium you may find that by using such a dense rock, you displace more water than is really desirable.

34 Make sure the rocks you use are free of pollution

Bear in mind that any rock stored outdoors may well have been exposed to all manner of pollution, including pesticides, herbicides and petroleum products, which may not be readily detectable. A thorough scrubbing before use would deal with most external contamination. Old rock from a discarded tank, if not cleaned at the time the tank was disassembled, could also represent a source of pollution.

35 Recycled rock may contain traces of copper

In a reef tank, beware of recycling rock from a fish system where copper medication may have been used. It is quite possible for the rock to have absorbed copper, which in time can leach out to the detriment of your invertebrates. If in doubt, soak the rock in saltwater and then check it using a copper test kit. If you are still in doubt, use a commercially available copper-absorbing medium as needed.

36 Artificial anemones can provide security

If you are maintaining any clownfish in a fish-only aquarium, try giving them an artificial anemone. You can buy imitation anemones, made out of silicone rubber (or other flexible materials), which a clown may adopt, helping it to feel more secure if it is in the company of larger fish.

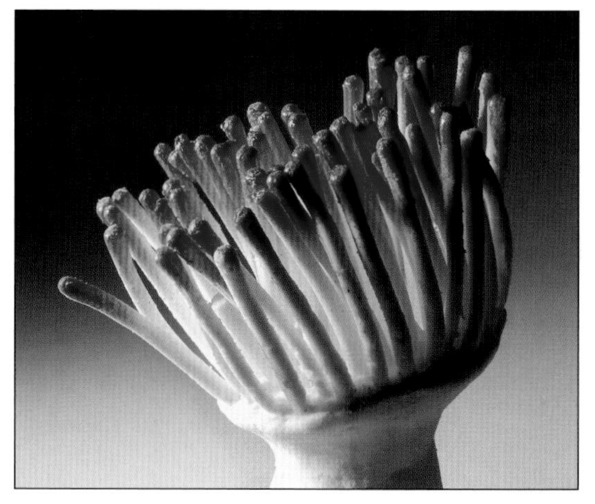

***Tip 36** An artificial anemone is suitable for a fish-only system.*

37 Imitation corals and sponges can add variety

Many manufacturers produce very realistic imitation corals and sponges. These can be used to decorate fish-only systems and to provide a bit of extra color. Coral reproductions with an elongate, tree or branchlike structure contrast well with rockwork, which tends to have more of a low, moundlike form.

Tip 37

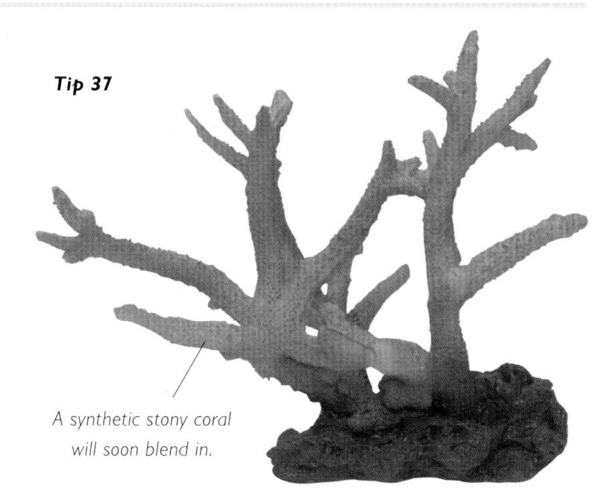

A synthetic stony coral will soon blend in.

38 What is live rock and why should I use it?

Live rock is mainly coral formed as a result of storm damage. Usefully for aquarists, it can be considered a natural, renewable resource. It tends to be highly porous and is thus capable of supporting large populations of bacteria, along with many other small organisms, such as algae, crustaceans, molluscs, sea squirts, sponges, worms, etc. Live rock introduces animals that help to turn the captive reef into a diverse, natural system, while its bacterial component is more than capable of processing the organic wastes produced by the larger inhabitants of the aquarium.

This synthetic sea fan looks surprisingly realistic and contrasts with more upright shapes.

Tip 39 *Live rock is functional as well as decorative.*

39 Choosing live rock with good potential

Look for low-density rock, preferably pieces showing signs of interesting life, such as forams, calcareous algae, tube worms, macro-algae, bivalves, etc. Do not expect to find a whole bunch of corals on this rock; any corals you do get are a bonus. What you are looking for at this stage is rock with the potential to bring useful life into your reef and support the prerequisite bacterial population that will help to process the waste products from the animals.

Better Aquarium Furnishings

40 Do I have to add all the live rock at the same time?

If finances are short and you are trying to build a reef on a budget, then you are better off starting with a limited quantity of good-quality live rock, enough to give you a biological kick-start, and then building up the remainder of your reef with well-cured rock as you can afford it. You do not have to put in all the rock at once. You could, for example, start with a coral outcrop ("bommie") at one end of the tank, with the intention of building toward the far end at a later date.

41 Creatures in live rock could cause problems in the aquarium

Be prepared to see all manner of creatures emanating from live rock. Not all may be friendly. Quite often, predators such as mantis shrimp may be lurking within holes in the rock. (See also Tips 296–312.)

42 Introducing live rock into the aquarium

Live rock should only be introduced into the aquarium when you are confident that both the temperature and salinity of the water are stable. Depending on tank size, this could be within 12 hours of the water being mixed. When introducing live rock to the aquarium, try to equalize the temperature by placing the bags containing the rock in the aquarium for 15 to 20 minutes. Acclimatizing the rock to the salinity of your water is a little harder, as it is likely that there will be little or no water in the bags, so just add water a little at a time until the rock is submerged. Remove the rocks from the bags underwater, and then rotate them, agitating them all the time, in an effort to expel as much air from the rocks as possible. By doing this you help to minimize secondary die-off of the small life forms resident in the rock. If possible, repeat this agitation a few hours later for the best results. (See also Tip 208.)

Tip 43 *Leave space in the aquarium for corals as well as live rock.*

43 Do not fill the tank with rock—leave room for corals

If you are setting up a reef aquarium, bear in mind that the live rock is just the infrastructure that the reef will be built on. Remember to leave room for the corals you will be adding later on.

44 Buy a magnifying glass to examine live rock

An essential piece of equipment for the saltwater fishkeeper is a magnifying glass or jeweler's loupe. Without one, you will miss out on all the tiny creatures found on live rock.

45 Arranging live rock in the aquarium

Arranging live rock is a skill that cannot be taught, but as a general guide, position the largest pieces at the bottom of the aquarium and then work upward, using incrementally smaller rocks. Leave spaces between the rocks to allow both the animals and the water free movement. In a smaller aquarium, choose pieces of rock in proportion to tank size. Beware of displacing your already limited water volume with excessive quantities of rock. Two approximately fist-sized rocks placed 2 inches (5 cm) apart will displace less water than a single rock of the same overall volume. As you arrange the rocks in the tank, keep stepping back and looking through the front glass to check on the progress of your design.

Tip 45 *Create a secure and attractive live rockscape.*

Tip 47 *Secure rocks from behind to disguise the joins.*

Be sure to build up rocks securely for a stable display

When placing rockwork into position, test each rock for stability before moving on to the next. Make sure that the rock is stable and unable to move, preferably with the center of gravity weighted toward the back of the tank. If the rock you are trying to position won't "lock" then try another. Putting rockwork together is much like doing a jigsaw puzzle without a picture, but with practice you will feel when you have done it correctly.

47 There are various methods of adhering rock to make a reef

Be warned that rocks are easily toppled by the burrowing actions of crabs and shrimps or are simply dislodged by crabs, snails and sea urchins determined to retrieve food particles from between the rocks. There are many different methods of arranging and adhering rockwork, ranging from gluing the rock together, through building concealed platforms out of glass or "egg crate," or drilling the rock and then using acrylic rods as retaining pins.

Better Aquarium Furnishings

48 Reef racks help you to build up the rockwork

There are a few commercially available items that can help you to arrange or position rockwork. A number of manufacturers produce reef racks that slot together to make tiers with three or four shelves on which to arrange corals or rock. One company produces a reef holder that clamps to the edge of the aquarium or to one of the strengthening bars. This allows you to extend rockwork up to the water surface or to suspend the rock to form caves.

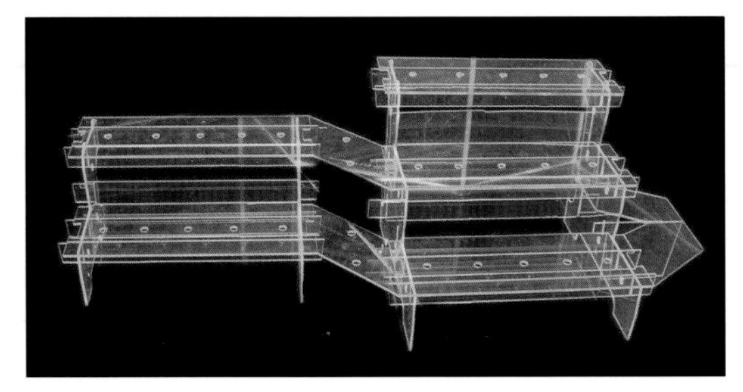

Tip 48 *Reef racks allow you to build up tiers of decor.*

49 Making the most of a limited amount of live rock

Racks can be used to good advantage in smaller aquariums. The smaller the aquarium the more water is going to be displaced by the rockwork, so by using a rack you can have a tank that appears to have substantial rockwork without losing those precious extra few gallons of water. A rack system can also be useful if you are incorporating live rock but are working with a limited budget.

50 Leave space for the fish in the display aquarium

An undersea cliff face looks impressive, but do remember to allow enough free-swimming space for the fish. If you do install such a rock feature, ensure that no fish can be trapped behind it.

51 It is possible to construct your own reef racks

You can easily make your own reef racks using acrylic sheeting or "egg crate." Egg crate is a plastic grid used as a diffuser below fluorescent lighting and is obtainable through aquarium dealers or from electrical wholesalers. Make sure you use material made from a food-safe plastic. Egg crate can be tied together using cable ties.

Tip 50 *Cliff faces are dramatic, but must be safe for fish.*

52 How important is the depth of substrate in a saltwater aquarium?

The answer: not as important as it used to be. The reason is that the traditional subgravel filtration system has been superseded by more modern methods, so a mere decorative covering is really all that is needed. However, in some instances a reasonable depth is still required, such as when the Jaubert/plenum or deep sand bed (DSB) system is used, and where burrowing fish such as wrasse or yellow-headed jawfish *(Opistognathus aurifrons)* are kept. (See also Tip 110.)

53 Burrowing fish need substrate materials in a range of sizes

If you plan to keep animals that burrow in the substrate, it is important to take the composition of the substrate into consideration as well as the depth. Whereas a sand-dwelling anemone will be satisfied with 4 inches (10 cm) of coral sand, an animal that actively tunnels, such as a pistol shrimp (*Alpheus* spp.) or jawfish, will need a mixture of particle sizes from sand, through gravel to rubble. This allows their burrowing to be shored up to prevent collapse.

Tip 54 *Coral sand is a suitable substrate material.*

Eliminating ferrous metal from sand substrate

It may not be necessary to wash the sand before adding it to the tank, but you may want to run a magnet over it first in case there are any particles of ferrous metal present. When all the sand is in, use a fine net to fish out any floating bits of organic material.

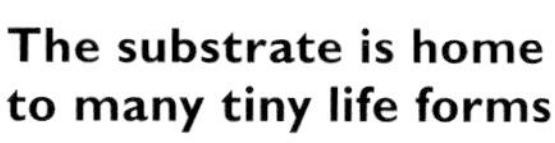

The substrate is home to many tiny life forms

About an hour after the aquarium lights turn off, sneak back quietly with a flashlight and examine the surface of the substrate. You will find it teeming with activity as tiny bugs go about their nocturnal business.

Tip 53 *The yellow-headed jawfish* (Opistognathus aurifrons) *sits vertically in its burrow.*

Better Heating and Lighting

56 What size heaterstat should I use in a fish-only aquarium?

Usually, a general guide is to allow 10 watts per gallon (or 2 watts per liter), with the total wattage required being split between two units where large tanks are concerned. Make any adjustments to the heater's thermostat control by very small amounts—an eighth of a turn is plenty to start with. Be sure to allow 20–30 minutes to elapse before rechecking the result.

57 Don't take any chances—use two heaters in the aquarium

It is good practice to divide the required heating between two heaters. This means that in the event of one heater's thermostat sticking in the "on" position, the risk of the aquarium overheating—and the consequential loss of valuable livestock—is drastically reduced.

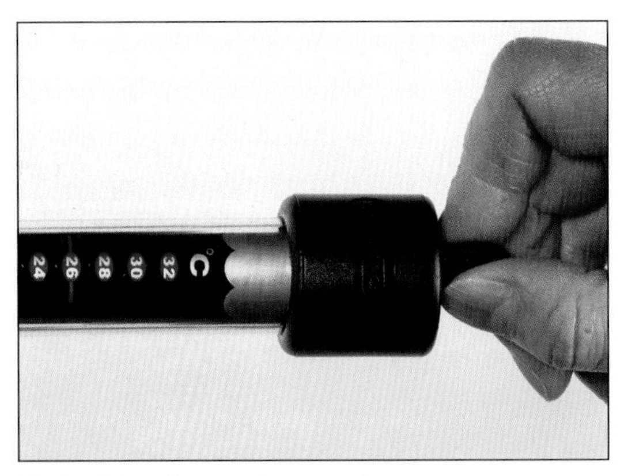

Tip 56 *Make adjustments in very small amounts.*

58 Fit heater guards to prevent injury to fish and invertebrates

As many fish enjoy resting on underwater obstacles, fit a guard over the heater to prevent them from burning themselves. This is even more important with slow-moving animals, such as anemones, echinoderms and molluscs.

59 Temperature control in the reef aquarium

In a reef aquarium of moderate size, say 53 gallons (200 L) or more, heating is not that big an issue. Heaters are required to bring the water up to temperature when you first start the reef off, but subsequently, in the average domestic setting, they will be doing very little work except on the coldest of nights. Think of your aquarium as a storage heater; while it is lit during the day, the heat from the lights is transferred to the water, where it is stored.

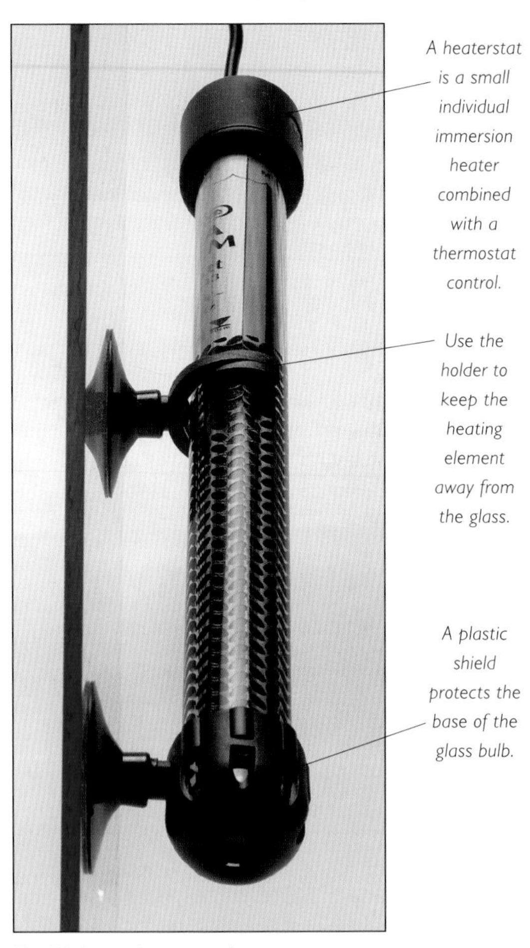

Tip 56 *A typical aquarium heaterstat.*

Tip 59 *Large reef aquariums do not need a heater.*

60 Allow heaters to cool before removing them

When removing a heater from the aquarium for cleaning, always switch it off and allow several minutes to pass so that any residual heat can dissipate. A hot heater is very easily dropped—usually onto a hard surface with predictable results!

61 Long-term temperature control in a reef aquarium

Because of the amount of lighting used above a reef tank to illuminate the corals, it is more likely that you will encounter problems in keeping the temperature below the maximum tolerated by tropical saltwater organisms. This temperature may be variable in captivity and will, to an extent, be dictated by the usual running temperature of the system. A rapid rise in temperature of several degrees or sustained temperatures above the 86°F (30°C) mark can cause coral bleaching. Depending on the species, lethal temperatures can start as low as 90°F (32°C).

62 Use more than one external thermometer

By using a number of external stick-on liquid crystal display thermometers fixed at varying depths, you can check on water temperatures around the aquarium. Adjust the position of powerheads to eliminate any cold areas.

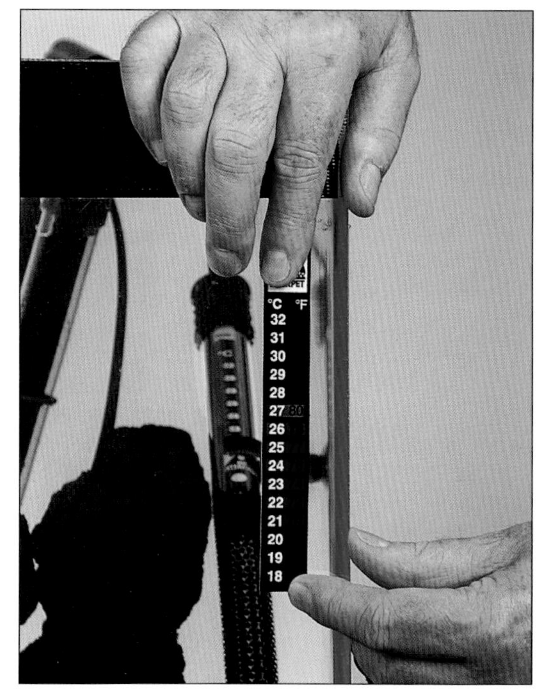

Tip 62 *Monitor temperature with an external thermometer.*

63 Should I buy a cooler for temperature control?

Installing a cooler is the sophisticated way of cooling the aquarium water. Some aquarists feel that in temperate climates it would be hard to justify the expense, but given the value of livestock that a reef tank can contain, plus a moral responsibility to the animals, a cooler is easily justified. Overheating is becoming more and more common, especially with the trend to increased levels of lighting.

Better Heating and Lighting

64 What are the options if I buy an aquarium cooler?

There are three main designs of cooler: those based on the use of refrigerant gas, on evaporative cooling and on the Peltier element. Each has a different rate of energy consumption. Those based around the Peltier element (a thermoelectric system) are the cheapest coolers, but the least efficient. Generally speaking, they are only suitable for smaller tanks.

65 Evaporative coolers are a cost-effective option

Evaporative coolers are the next most expensive. They are efficient and economical to run, but they will cause a great deal of additional evaporation, which means that you will definitely need to run an auto top-up system in conjunction with them. And bear in mind that as a consequence of the additional evaporation, you will experience an increase in room humidity, so they are best vented to the outdoors to reduce this.

Tip 65 *An evaporative cooler*

66 Some refrigerant coolers have additional features

Refrigerant coolers are probably the best choice. They are highly efficient and easy to plumb in, but they are the most expensive option. Some of these units may have additional features. For example, they may also act as a heater or incorporate a built-in UV sterilizer.

Tip 66 *A refrigerant cooler is easy to connect to a saltwater aquarium.*

67 Cooling the display aquarium in an emergency

At a basic level, it is possible to achieve some cooling effect by floating plastic bags or a food-safe container of ice in the aquarium. A possible alternative is to freeze the reverse osmosis (R.O.) water you use for topping up evaporation losses and then allow it to melt in the tank. These strategies are more suited to emergencies than as a long-term solution.

Tip 67 *Ice cubes in a plastic bag are an emergency measure.*

Tip 64 *Strongly lit reef aquariums may need coolers to maintain the temperature at acceptable levels.*

68 Evaporative cooling using an electric fan

You can also create a system of evaporative cooling by directing the airflow from an electric fan across the surface of the tank or sump. Remember that you will need to top up with greater amounts of freshwater. The type pictured is more common in Europe, but a variety are available in North America. If you use a fan close to your tank, make sure it is firmly secured so there is no danger of it falling into the water or getting wet. Also bear in mind that the fan may display signs of corrosion after a time and will probably need replacing after a year or two.

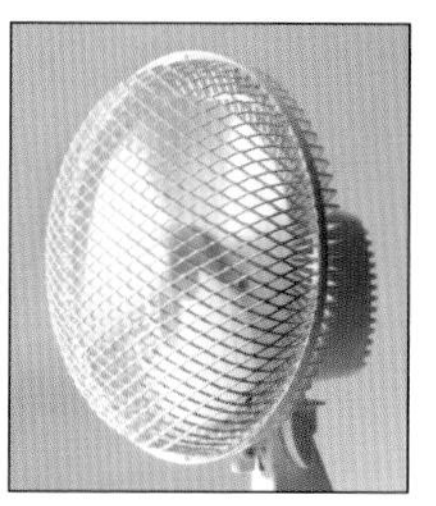

69 An extractor fan can also perform a cooling function

If your aquarium is against an outside wall, a useful cooling solution is to install an extractor fan that will suck warm air from above the surface of the water and blow it out of the room. An extractor fan should be balanced with ventilation to allow fresh air to enter the room, preferably from a cooler location.

70 Will air conditioning help to keep my aquarium cool?

Air conditioning can be a great solution to prevent a tank overheating in the summer, with the added bonus of keeping the reefkeeper cool as well! Unfortunately, air conditioning can be expensive, both to buy and to install. Portable air conditioners are cheaper, but the running costs may be quite high unless they have a good thermostatic control.

71 Energy-saving solutions for cooling systems

You can improve your energy usage of fans, etc., in a couple of ways. One is to control them with a timer so they come on shortly after the metal-halides fire up. This is an easy solution for those times when you are out of the house. Alternatively, use a thermostatic cooler controller. Set the controller to the desired temperature and whenever that temperature is reached it will turn on the cooling equipment. When the fan or air conditioner has done its job of lowering the temperature, the controller turns it off. This type of controller can also be used to regulate the room temperature of a multitank facility by using the sensor to read air temperature, rather than individual tank temperature.

Better Heating and Lighting

72 Do you want your fish to look colorful or natural?

The appearance of a fish's color depends upon the "color" of the lighting used in the aquarium. The color tone of any light source is classified in terms of its temperature and expressed in degrees Kelvin (K). A candle flame is a very "warm" light source with a low color temperature of about 1,800 K, whereas midday sunlight in the tropics is about 6,500 K. Lamps running at higher color temperatures (e.g., 10,000 K, 14,000 K and even up to 20,000 K and higher) give out a harsher, colder, bluer light as the color temperature increases. "Warm" lights make reds and yellows look good, whereas the colder lights give the fish a much paler rendition. However, it must be said that these are the colors you would actually perceive in the sea, where only the blue end of the light spectrum penetrates to any great depth.

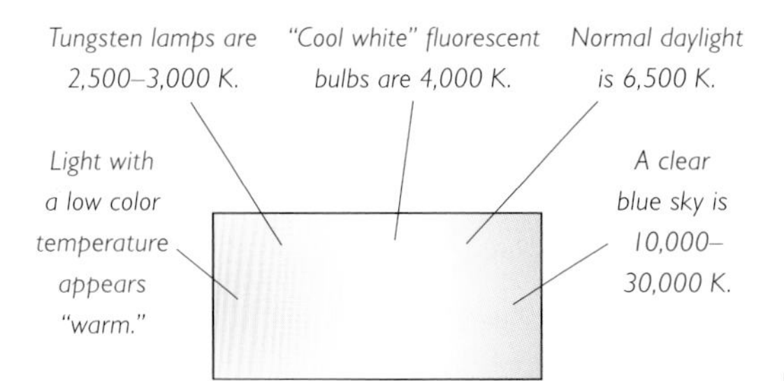

73 What type of fluorescent bulbs are available?

There are three main types of fluorescent bulbs available for the aquarium hobby. The traditional T8 bulbs are 1 inch (25 mm) in diameter and are also widely used in homes and offices. They are available in various colors. The T5 bulbs are 5/8 inch (16 mm) in diameter and provide a brighter light output for their size. These are also available in a compact PL format, in which the bulb is bent back on itself and has four connecting pins at one end of the fitting.

74 Maximize the effect of light with clip-on reflectors

Unless the interior of the aquarium hood is lined with reflective material, much of the light generated by the fluorescent bulbs is wasted, going upward rather than down. Clip-on reflectors, which almost totally surround the top half of the bulb, help to maximize the effect of light generated by fluorescent bulbs by directing the majority of it down into the water.

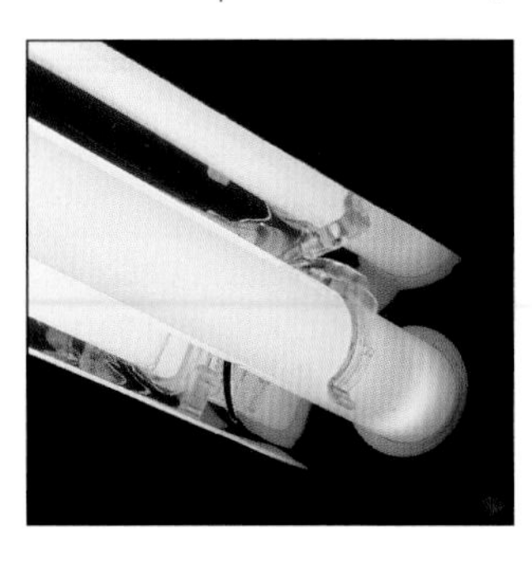

Tip 74 *Curved reflectors maximize light from bulbs.*

75 What are the pros and cons of using metal-halide lamps?

Metal-halide lighting units provide a bright light that simulates natural sunlight, giving the reef aquarium glitter lines, highlights and shadows, as they would appear in nature. They are expensive to buy, however, and need to be suspended above the aquarium to provide ventilation for the heat they generate.

Tip 73 *Three fluorescent bulb types, from top: T8, T5 and PL.*

76 How do I choose between fluorescent and metal-halide lighting?

Fluorescent lighting is really only suitable for tanks up to 18 inches (45 cm) deep, and metal-halide lighting is essential for deeper reef aquariums. For general, soft-coral-based reefs up to 24 inches (60 cm) deep you can achieve good results with 150-watt bulbs, although results would be even better with 250-watt bulbs. For reefs based around more light-demanding corals, such as *Acropora* species, start with 250-watt bulbs, but consider 400-watt bulbs, especially if the depth of the tank is greater than 24 inches (60 cm).

77 Keep everything clean for maximum light penetration

Even well-directed light can be wasted if the cover glass is dirty or there is a lot of suspended debris. Biofilms or surface films also cut down light. Keep cover glasses spotlessly clean and maintain mechanical filters regularly.

Tip 77 *Maximize your lighting.*

78 Corals need time to adjust to more powerful lighting levels

More exotic corals require much higher lighting levels than others. Be aware, however, that some corals can be adversely affected by sudden exposure to more powerful lighting. To make the transition less stressful for them, suspend a metal-halide pendant light further above the tank than it would ideally be, and gradually lower it to its final position over a period of several days. This will enable the tank inmates to become accustomed to the increase in light levels.

Tip 79 *White and actinic blue marine bulbs.*

79 What is the advantage of using actinic bulbs?

Actinic bulbs give out light that peaks in the blue part of the spectrum (at 420 nm). This light is particularly appreciated by invertebrates, including corals. Actinic bulbs are often fitted in conjunction with whiter bulbs and switched on before (and switched off after) the white bulbs to simulate dawn and dusk lighting. These periods hardly exist in nature, but creating a smoother passage between bright light and darkness reduces the stress risk to the fish.

Better Heating and Lighting

80 See corals in a new light as their colors fluoresce

During your dusk period when your reef is lit solely by blue actinic light, you can enjoy a stunning display as the colors of many of the corals fluoresce. Often, they will exhibit different coloration than under white light. Some corals of differing species that appear to be the same color in regular light, are now seen to have completely different colors from each other.

81 How many bulbs do I need in a saltwater aquarium?

For a reef aquarium, use a minimum of four bulbs on tanks up to 15 inches (38 cm) wide (front to back measurement). Larger tanks will need more bulbs. Fluorescents will not provide enough light for tanks more than 18 inches (45 cm) deep. If you are using metal-halide units, one will illuminate a tank up to 39 inches (1 m) wide. Otherwise, calculate one lighting unit for every 24 inches (60 cm) of tank length.

Tip 81 *Metal-halide and fluorescent bulbs provide excellent light.*

82 Obtaining the best effect from fluorescent bulbs

If you are installing a number of different color temperature fluorescent bulbs in the hood of a fish-only aquarium, put the actinic blue bulb at the back. White bulbs then light up the foreground of the display and the dimmer blue light gives a distance-enhancing effect, helping to make the tank appear deeper front to back. In a reef aquarium, position the actinic bulbs directly above corals that need enhanced levels of blue light.

Tip 82 *A combination of actinic and white bulbs.*

83 Simulate the length of a tropical day

Animals in tropical seas are creatures of photoperiodic habit. This means they are accustomed to 12 hours of daylight each day, with only a little graduation between daylight and darkness. Simulate this natural cycle with your aquarium lights. If using fluorescent lights you may be able to compensate for under-lighting to a certain extent by extending the photoperiod to 14 hours.

Light in the aquarium

Aquarium lighting should be similar to natural light, but can be altered slightly for practical purposes.

An actinic blue bulb coming on before the others provides a dawn effect each day.

At the end of the day, the white bulbs go out before the blue actinic one, creating a dusk effect.

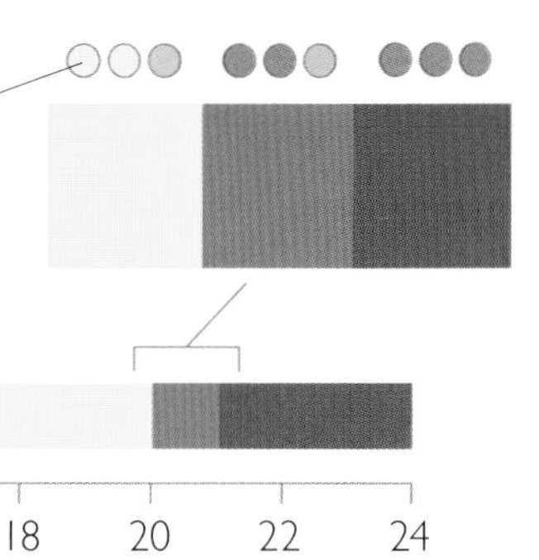

0 2 4 6 8 10 12 14 16 18 20 22 24

84 Use multiple timers to control aquarium lights

Always use timers to turn the lights on and off. At the very least, use separate timers for blue and white lights. If you run multiple white lights you may want them on individual timers as well. If you intend to use blue lights to provide a dusk/dawn effect, set the timer to turn them on 30–60 minutes before the white lights come on, and turn them off 30–60 minutes after the white lights go out.

85 Lighting levels in the early days of your aquarium

For the first one to two weeks of your reef aquarium's life, light the tank for around six hours a day. At this point you do not need to utilize the full complement of lighting that your finished reef will need; in fact, it would be to your disadvantage to do so. Because your reef is vulnerable to outbreaks of pest algae during its early days, it makes sense to limit the amount of light available until you are ready to start stocking with photosynthetic organisms.

Increase the photoperiod gradually and monitor results

As your reef starts to mature and you build up your stock of corals and other photosynthetic animals, gradually increase the photoperiod by half an hour to an hour, every four or five days, until you reach your desired day length. Keep a close watch for any signs of pest algae starting to take hold. If you see a problem with unwanted algae, adjust the lighting back by a couple of increments.

Adjust the lighting period to suit your viewing time

As far as deciding what time of day to have your reef's day starting and finishing, it's a good idea to work back from your bedtime so that the reef is lit when you are there to enjoy it.

Tip 87 *Time your lighting to suit your viewing.*

The typical tropical day is lit for about 12 hours from 7 a.m. to 7 p.m.

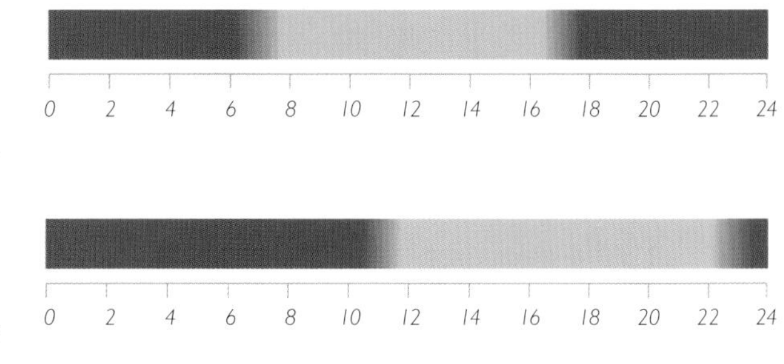

Shifting that period to 11 a.m. to 11 p.m. provides evening viewing.

Better Water Management

88 Maintaining stable water conditions is vital

The physical conditions found in coral reefs, such as the chemical constituents of the seawater, the temperature, salinity and specific gravity, are all extremely stable. Ensuring that you mimic this stability in your aquarium is essential for the well-being of the fish and invertebrates you introduce.

Tip 88 *A reef edge at low tide in Indonesia.*

89 Filtration is just one factor to consider

The choice of filtration should not become the only decision to warrant serious thought. There are many other factors that could ultimately cause the beginner much greater grief, such as poor understanding of the biological processes, mixing incompatible animals or allowing the aquarium environment to become unstable. There are many filtration systems to choose from, but if correctly set up and maintained, any biological filter should do the job it was designed for, admittedly some more efficiently than others.

90 Filling the tank for the first time

When setting up a new marine tank from scratch you can use it as a handy container in which to mix its own saltwater. You can do this ahead of putting in any rockwork, in which case don't fill the tank with water to its full capacity—when you put the rocks in, it will overflow!

91 Using R.O. water gets the aquarium off to a good start

For the reef aquarium, it is particularly important to make up your saltwater using freshwater containing as few nutrients as possible. It is best to exclude undesirable nutrients, including phosphates, nitrates and silicates, all of which can help fuel outbreaks of pest algae that can smother, overgrow or chemically affect the corals you want to keep. By limiting these from the very start, you will have a far better chance of keeping a reef without constantly battling against algae. At the very least, pass domestic tap water through an aquarium nitrate-removing resin. However, reverse osmosis (R.O.) filtered water is one of the easiest ways to obtain water of the quality required. You can treat water by passing it through your own R.O. unit. Alternatively, most saltwater aquatic dealers will have R.O. water for sale and will fill up your container for you.

Tip 91 *A reverse osmosis (R.O.) unit removes contaminants from tap water.*

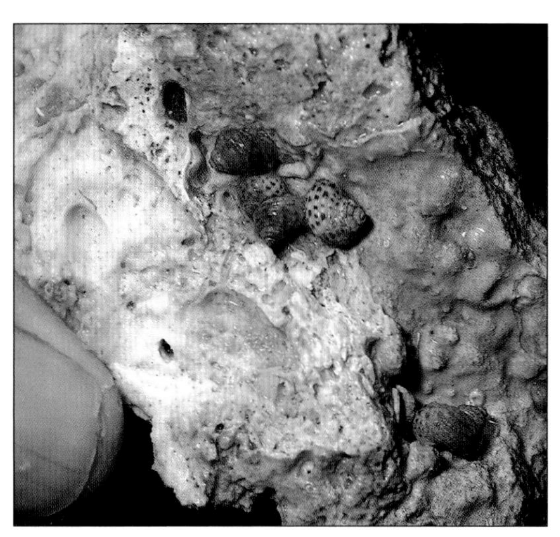

Tip 92 *Turbo snails (Turbo spp.) will deal with algae.*

92 Hermit crabs and snails help to prevent algae buildup

When setting up a new fish-only aquarium, allow the filtration system to mature (ammonia and nitrite readings should be zero) before adding any fish. However, running the tank with the lights on will encourage some algae to grow. To keep this growth down, introduce a "cleaner gang" of hermit crabs (*Dardanus* spp.) and snails during the maturation period. The hermit crabs are good for clearing up subsequent debris, while snails (*Turbo* or *Astraea* spp.) will eat up algae.

93 The new aquarium is a harsh environment

In the early days of a fish-only aquarium the environment is relatively harsh. The biological filter will not be at peak capacity and therefore will be susceptible to overload. Naturally, the inexperience of a beginner increases the possibility of mistakes. Extra vigilance is required at this time, as the combination of an immature tank and an inexperienced hobbyist can lead to a stressful environment, with the potential for disease outbreaks.

94 Adding bacterial cultures speeds up maturation

You can accelerate the aquarium maturation process in a fish-only system by adding special bacterial cultures that seed the filtration system. These are available from your dealer and, used properly, will help to establish colonies of nitrifying bacteria in a much shorter time.

95 Incorporate a drip-through trickle filter tray

If room exists, you can build a simple, drip-through trickle filter or trickle tower above the sump in a fish-only system. It is a very efficient form of biological filtration, providing maximum oxygenation and superior nitrification compared to submersed biological filters. It also has the advantage of suffering fewer problems during power failures than submersed biofilters. Trickle filters cause increased evaporation and, consequently, heat loss, which can be an advantage or disadvantage depending on the requirements of your system.

Trickle filter

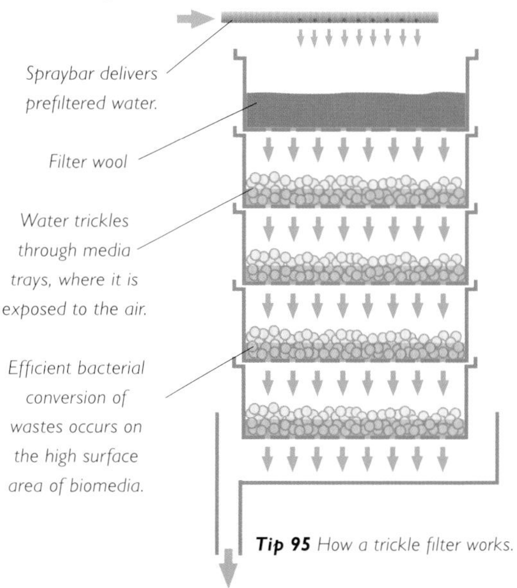

Tip 95 *How a trickle filter works.*

96 A fluidized bed filter is highly efficient

A fluidized bed filter used in a fish-only system has an amazing nitrifying capacity, but the efficiency of the bacteria involved seriously depletes the oxygen content of the water flowing through the unit. Always aerate the water as it returns to the aquarium, either by introducing air into the return bulb or using a spraybar to distribute the water over the surface. (Be aware that fluidized beds and biological filtration in canister filters are unsuitable for a reef aquarium.)

Tip 96 *A fluidized bed filter*

97 Isolating taps simplify cleaning an external filter

Install isolating taps on any external filter connected to a fish-only aquarium. For sheer practicality, they make filter cleaning simple and, more importantly, mess-free! As long as you fill the filter body almost full with water after cleaning, there will be very little air to purge from the system once you reconnect the taps and open them.

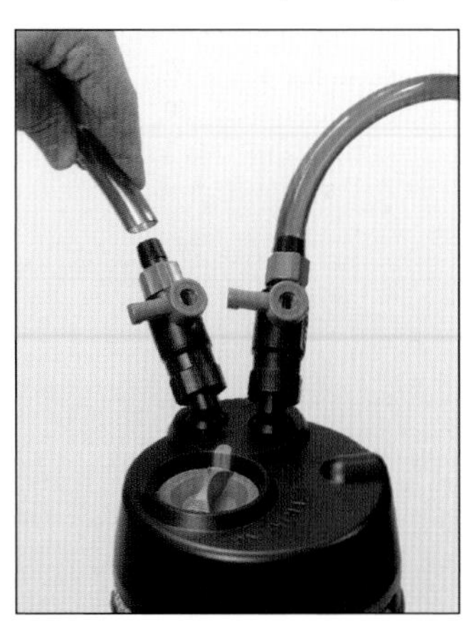

Tip 97 *Isolating taps simplify the filter-cleaning process.*

98 Inlet and outlet hoses can help to create water flow

Install the inlet and outlet hoses of the external filter at opposite ends of a fish-only aquarium to create a flow of water right through the tank. Reef aquariums should have more powerful and varied forms of water circulation. (See also Tips 134 and 135.)

99 Restoring pH values at the end of maturation

Normal seawater and new mixed saltwater (at the correct salinity, or specific gravity, and temperature) have a pH value of 8.3. At the end of the maturation period in a fish-only system, it is likely that the pH will have dropped toward 8.0 or below. A falling pH—water becoming more acidic—indicates that the water quality is deteriorating and that a water change is necessary. Carry out a 25 percent water change to restore the pH to its normal value.

100 Low pH can result from lack of water movement

Insufficient water movement can also be a cause of low pH readings. If there is not enough disturbance at the surface of the aquarium, CO_2 levels can build up in the water, depressing its pH. By increasing surface water movement, the excess CO_2 will be gassed off, returning the pH to normal. If you cannot see the surface rippling, then you most likely do not have enough flow.

101 Lack of ventilation can also contribute to low pH levels

Surface biofilms, or cover glasses that fit too tightly without allowing for ventilation, can be the cause of low pH readings, because the CO_2 present in the water (as a result of biological processes) is not allowed to escape. Be aware that a large gathering of people in a poorly ventilated room can contribute to decreased pH in your aquarium by putting unusually high levels of CO_2 into the room's atmosphere.

102 How to tell if a low pH level is caused by excess CO_2

You can easily test whether low pH is caused by excess CO_2. Measure the pH level of your aquarium, taking note of the reading, then immediately take a sample of the water—about 1/4 gallon (1 L) should do. Remove the sample to another room, aerate it for 12–24 hours and then measure the pH of the test sample. If the pH has risen, this shows that excess CO_2 has been gassed off in the intervening time, indicating that an accumulation of CO_2 is the cause of the pH problem in your tank.

103 Test the water at the same time each day

In a reef aquarium, pH varies over the course of 24 hours due to different biological processes taking place during the night and day. An aquarium will be at its lowest pH first thing in the morning, just before the lights go on, and at its highest at night when lights are turned off. To make meaningful comparisons of pH changes, always test at the same time each day.

How pH changes over 24 hours

A *Take a test at the end of the day to record the highest pH reading.*

During daylight hours, strong illumination promotes photosynthesis by corals and algae. Carbon dioxide (CO_2) is consumed during photosynthesis, resulting in an increase in the pH level.

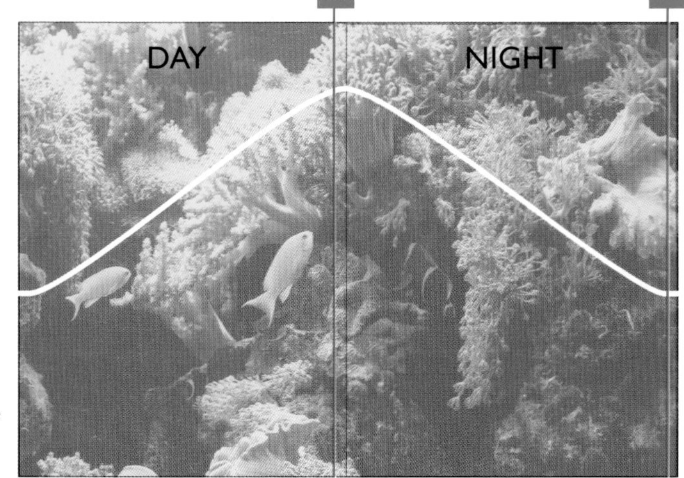

B *Take a test at the end of the night to record the lowest pH reading.*

At night, when there is no photosynthesis taking place, plants give off CO_2 through respiration. As CO_2 dissolves in seawater it forms carbonic acid, which lowers the pH level.

Tip 103 *The pH levels in a reef aquarium vary over the course of a day.*

Better Water Management

104 Carbon—a valuable treatment in a reef system

Carbon can be very useful in the reef tank. You can use it to remove the toxic compounds produced by soft corals (for the purpose of chemical warfare against rival species) and to clear up any discoloration of aquarium water. Even if you do not use it on a regular basis, it is well worth keeping some on hand to use in the event of any emergency where the aquarium water has become polluted.

105 Three strategies for using carbon in a reef tank

When it comes to using carbon in a reef tank, there are three common strategies. You can use it for 24 hours once a month; for seven days once a month; you can use it continuously, replacing it at regular intervals, preferably every six weeks or less. If carbon is used for extended periods it will start to perform as a biological filter, with the resultant risk of elevated nitrates. If your water is very discolored, do a water change first and then add the carbon. Be wary in this kind of situation, as sudden increases in the amount of light reaching the corals can cause bleaching.

106 Replenishing carbon in a fish-only system

In a fish-only system, replenish activated carbon filter medium regularly. If left for too long a period, it is likely to unload all the adsorbed materials back into the aquarium water. If still "active," a small portion of activated carbon will remove any yellow color from a sample of the aquarium water. Try testing some carbon after a month or so to see if it is still capable of doing its job.

107 Forget traditional filtration in a reef aquarium

Traditional methods of man-made biological filtration are neither advisable nor necessary in a reef aquarium. Man-made biological filtration processes ammonia to nitrite, then nitrite to nitrate, but stops there. This buildup of nitrate will act as a fertilizer for pest algae. Forget about filtration—think more in terms of water treatment. At the very simplest level, all that is needed for a reef aquarium to work is lighting and water flow appropriate to the species being kept. Live rock, mud or deep sand beds are the best ways of completing the nitrogen cycle in the reef aquarium.

Nature's nitrogen cycle

Denitrifying bacteria reduce nitrate to nitrogen gas.

Fish digest and metabolize protein and excrete ammonia.

Nitrite is converted into nitrate by nitrifying bacteria.

Ammonia is converted to nitrite by nitrifying bacteria.

108 Live rock takes care of the maturing process

When using live rock as the basis of a reef aquarium, there is no need to add anything for the maturation process. The bioload of the live rock will take care of that itself.

109 The Berlin method of aquarium filtration

The Berlin method combines protein skimming and the corresponding nitrifying and denitrifying actions of bacteria on the surface and internal interstices of live rock. It is much nearer to a "natural" method of water management in a reef aquarium compared to earlier methods using man-made biological filtration.

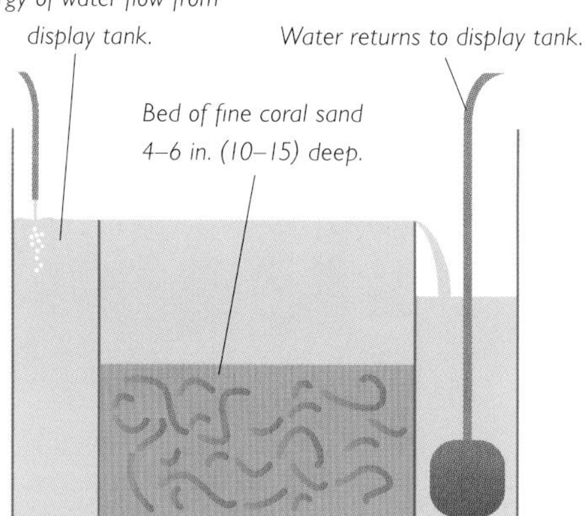

Tip 110 *A deep sand bed filter*

110 How does a deep sand bed (DSB) filter work?

The philosophy behind the deep sand bed (DSB), also called plenum filtration, as used in reef aquariums is very close to that of the Berlin system, but the emphasis here is on a sand bed, 4 inches (10 cm) deep or more, to do the natural filtration and denitrification work of live rock. In theory, the vast surface area offered by a DSB system will far exceed that offered by a conventional live rock system, allowing you to considerably reduce the tank volume taken up by rockwork. To work successfully, a DSB must be populated by plenty of small "creatures," or infauna, various worms and microcrustacea, plus meiofauna—predominantly the microbial colonies living on the sand. This is easy to achieve by seeding the bed with about 1 pint (0.5 L) of sand from an established reef.

Bioballs break up the flow of water from the display tank.

Bioballs prevent loose algae pieces being sucked into the pump.

Fluorescent bulb to illuminate macro-algae.

Water returns to display tank.

Water flows through gap in first partition.

Water pump

Substrate layer about 2 in. (5 cm) deep supporting growth of macro-algae.

Tip 111 *The mud system*

111 The mud system relies on macro-algae

The mud system, often called a refugarium, is a natural form of filtration for a reef aquarium that utilizes a fine substrate, lit 24 hours a day, where macro-algae, in the form of *Caulerpa*, are grown. It does not require a protein skimmer or activated carbon to purify the water. It keeps the water well oxygenated through the photosynthesis of the *Caulerpa*, which being lit 24 hours a day, helps to stabilize the pH. This method uses the end products of fish metabolism and fixes them into plant tissue, thus preventing the accumulation of ammonia, nitrites, nitrates and phosphates. The *Caulerpa* prevent the growth of undesirable hair algae by competing for essential nutrients; it can be physically harvested to remove nutrients from the system and can also be used as a fresh, natural food for fish.

112 Choose a good-quality protein skimmer

The efficiency of a protein skimmer is in direct proportion to the time that water is in contact with the microscopic air bubbles generated in the reaction chamber. The longer the path the water has to take, the better. Take a look at the protein skimmer's design before you buy. Relatively inexpensive air-operated systems may not incorporate long flow paths or even countercurrent exposure to the bubbles.

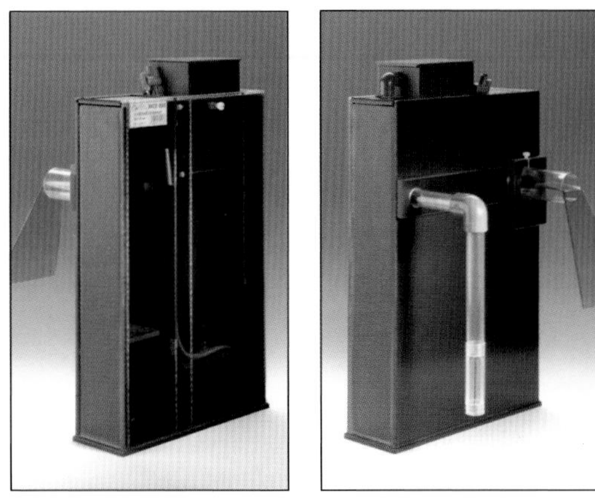

Tip 112 *Protein skimmer suitable for the display aquarium.*

113 The protein skimmer must produce concentrated foam

Adjust the rate of the protein skimmer to produce the correct type of foam. Large amounts of wet foam simply waste water; the aim should be to produce concentrated foam. The performance of the skimmer will vary with the bioload placed upon it through feeding, aquarium additives and extra livestock. If the skimmer suddenly bursts into an extra vigorous production of foam, this is a clear sign of an abnormal increase of waste organic matter in the tank, perhaps as a result of overfeeding or from a decomposing fish or invertebrate. Always investigate!

Tip 114 *A protein skimmer suitable for a sump.*

114 How does a protein skimmer work?

A protein skimmer works on the principle that molecules of organic waste, such as those produced in a saltwater aquarium, are attracted to the surface of air bubbles. In the skimmer's reaction chamber, a flow of aquarium water is in close contact with rising columns of air bubbles that carry the attached molecules upward toward a skimmer cup, where they collapse and form a yellowish liquid that can be removed. Water should be drawn into the skimmer from as near the surface as possible.

Tip 118 *For the best results, keep the nitrate level at zero.*

115 Do not allow ozone to escape into the atmosphere

The recommended method of using ozone safely in a fish-only system is to introduce it via the protein skimmer, thus isolating it from direct contact with the main body of aquarium water. To ensure that no ozone escapes into the atmosphere, fit an activated carbon filter on top of the skimmer's collection cup.

116 Match the capacity of the ozonizer to your tank

Ozone is potentially dangerous to the creatures in your aquarium, and to you and your family. Never install an ozonizer with a capacity greater than what is required for your tank. Ideally, install one where the maximum capacity is the same as the tank.

117 Choose safe materials when using ozone

Ozone is a well-known destroyer of many aquarium equipment materials, such as rubber and air line tubing. Never restrict ozone with standard air line valves—they will melt! Always use ozone-tolerant materials.

How much nitrate in the aquarium is too much?

This question is open to debate. Many fish are quite tolerant up to 150 mg per gallon (40 mg/L), if they have endured a gradual rise in nitrate levels, rather than been suddenly exposed to this high level. In a fish and invertebrate collection, it is generally recommended that you try to keep nitrate levels as low as possible, and most saltwater fishkeepers appear to achieve minimal levels, 0–4 mg per gallon (0–1 mg/L), without too much effort.

119 Why is it important to limit nitrate in a reef aquarium?

There are two main reasons for limiting nitrate in a reef aquarium. Firstly, nitrate acts as fertilizer for pest algae, which can go on to smother corals by overgrowing them or adversely affect them through allelopathy (chemical warfare). One of the main reasons why people leave the hobby is probably because they lose this battle and become disheartened. Secondly, nitrate can interfere in the process of calcification, slowing the growth of corals. It is worth noting that some animals, such as clams and soft corals, are able to use nitrate directly as a nutrient.

Better Water Management

120 Specific gravity and salinity, are they the same?

The terms specific gravity (S.G.) and salinity can cause confusion. Within the marine hobby, aquarists commonly use specific gravity, which is actually a measurement of density, as a means of assessing the saltiness of the water. Unfortunately, the correlation between specific gravity and salinity depends on temperature. This means that you would get different readings of S.G. from the same sample of water if you tested it over a range of different temperatures. Salinity is a measure of how much salt is in the water—usually expressed in terms of parts per thousand (ppt). This is the same as grams per gallon or liter—a much more useful and practical way of thinking about salinity.

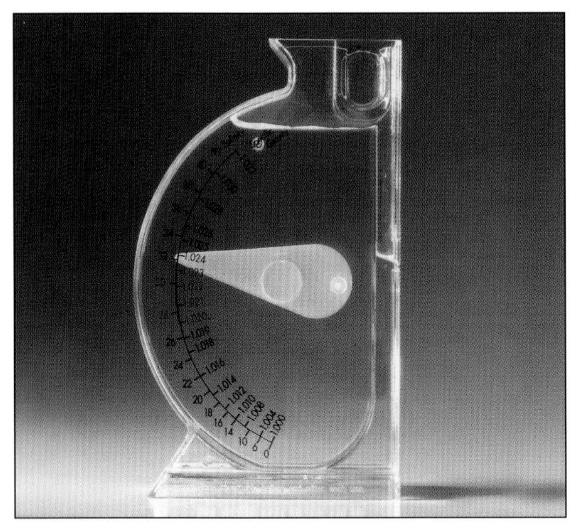

Tip 121 *A swing-arm hydrometer measures specific gravity.*

121 What is a swing-arm hydrometer used for?

Hydrometers measure the amount of salt in the water in terms of specific gravity (S.G.). A hydrometer will only give an accurate measurement within a narrow temperature range, and that range must be appropriate to the temperature of your tank.

122 What is the best salinity for my aquarium system?

For the reef tank, aim for natural seawater levels of salt, recognized as a salinity of 35 ppt (140 gm of salt per gallon of water, or 35 gm/L), or a S.G. of 1.023. The fish and corals have evolved for this level. Test kits may well be inaccurate when measuring certain parameters at unnatural salinity. Hyposalinity (lower than normal) should only be used as a therapeutic measure.

123 What is a refractometer and how does it work?

The refractometer is a better, slightly more expensive way of measuring salinity. It is easy to use, it automatically compensates for temperature, you can easily test and recalibrate it, and you can read off an accurate measurement of salinity in parts per thousand. Refractometers are optical instruments; they use a prism to measure the refractive index of a liquid. As a beam of light enters the sample of liquid, it is refracted, or bent, with the amount of refraction being proportional to the salinity of the sample. To use a refractometer, place a couple of drops of water on the prism, cover it with the transparent hinged plate, then look through the eyepiece to examine the reading. The bottom of the scale represents zero salinity; by using a sample of R.O. water as a test solution, the refractometer can be calibrated by means of an adjustment screw so that the indicator line intersects this point.

The edge of the blue area shows the salinity.

Place a water sample on the viewing window.

Tip 123 *A refractometer*

124 Electronic hand-held meters are accurate but expensive

Electronic hand-held meters are very accurate but far more expensive. These instruments give a readout in microsiemens (mS), necessitating the use of a conversion chart to translate the reading into salinity—in parts per thousand (ppt)—or to specific gravity. Fortunately, a reading of 53 mS translates to a salinity of 35 ppt—the natural seawater level. This is an easy figure to remember, as it is the same two digits transposed. Electronic meters can be recalibrated to maintain accuracy, and they automatically compensate for the temperature of the water being tested, thus ensuring an accurate reading.

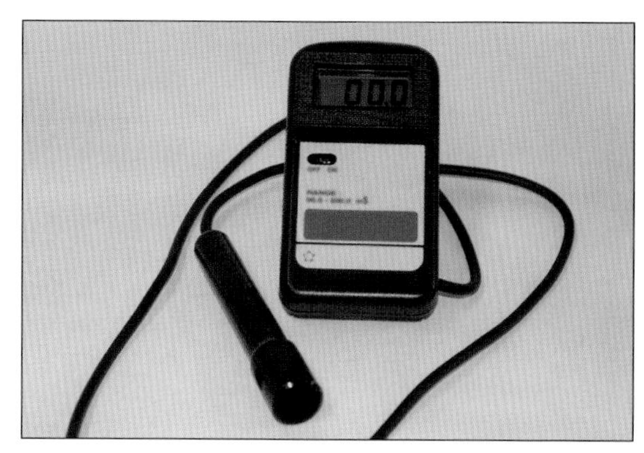

Tip 124 *An electronic meter gives an accurate reading.*

125 Keep instruments clean to maintain accuracy

Over a period of time, any instrument used to measure salinity can become inaccurate if subjected to occasional clumsy usage or if it is not cleaned properly. To clean a swing-arm hydrometer, periodically fill it with vinegar and leave it to soak overnight to remove any buildup of salt or carbonate deposits. Rinse the hydrometer with R.O. water following this procedure. Refractometers and electronic meters should be recalibrated on a regular basis according to the manufacturer's instructions.

Tip 122 *The natural salinity of seawater is 35 ppt.*

126 How should I prepare water for water changes?

Start by adding the required amount of R.O. water to a food-safe bucket. After about 15 minutes, turn on the heater (making sure it is fully immersed), and start agitating the water with an air pump or powerhead. When the water has reached a temperature of 77°F (25°C), add the salt— 140 gm per gallon (35 gm/L), i.e., 1,400 gm for 10 gallons of water (350 gm for 10 liters of water). Leave the solution to mix for 24 hours before using it. Be sure to turn off the heater and give the element a chance to cool before emptying the bucket.

127 Do not risk contamination when mixing saltwater

To keep the risk of contamination to a minimum, use food-quality containers, such as buckets, etc., when mixing saltwater. Buckets such as those used for winemaking are very suitable. Plastic containers are often marked with the symbol of a wine glass and a fork to signify that they are food-safe items.

Better Water Management

128 Seal your salt mixes to prevent water absorption

Salt mixes are strongly hygroscopic, meaning they will absorb water out of the atmosphere. If this happens, and you calculate the amount of salt needed for a given salinity in a given volume of water, your calculation will no longer be accurate due to the increased water content of the salt. So be sure to store salt in an airtight container and reseal it after use.

Check salinity regularly in the early days

Check the salinity of your water every few days when you start out, as your reef will experience a certain amount of evaporation. The amount will depend on a number of factors, such as ambient air temperature, whether your tank and/or sump are covered and what type of lighting you are using. You will experience a greater rate of evaporation during the summer when the air temperature is higher, if the tank or sump are open or if you use metal-halide lighting (or even T5 fluorescent bulbs, but to a lesser extent).

Maintaining a stable level of salinity

By putting a permanent mark on the side of your aquarium at the required water level at the required salinity, you will be able to see at a glance when an evaporation top-up is needed. If your system uses a sump tank, remember that the only place evaporation will be evident will be in the pump compartment of the sump, so make your mark there. The maximum level at this location will depend on the amount of water the sump can hold in the event of a power failure. Get this wrong or overfill the sump and a wet floor is inevitable. Today, given the low cost involved, it is well worth installing an auto top-up system to ensure that salinity remains stable.

131 Note the amount of salt required for water changes

After the first few water changes, you will get to know roughly how much salt you will need. Mark off the water level on the outside of the bucket, together with the weight of salt required to give the correct specific gravity. This will speed up making new salt mixes.

Auto top-up system

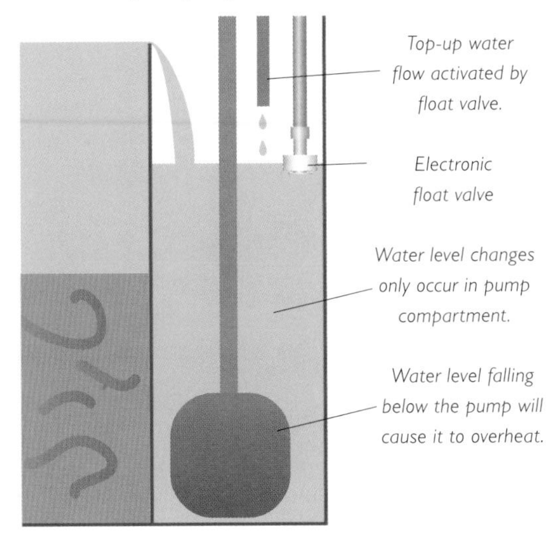

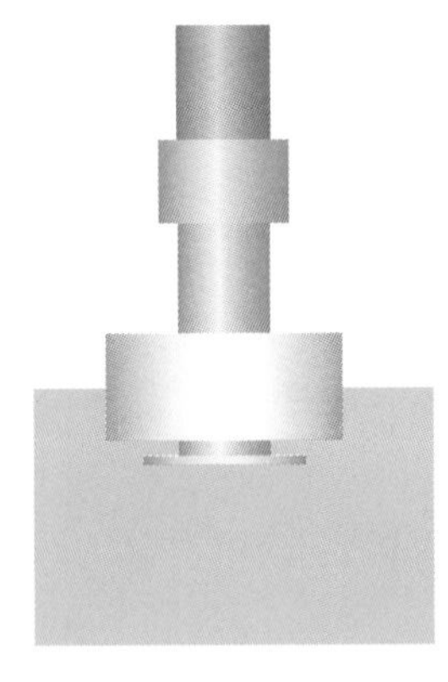

As water level drops, the float falls, switching on top-up pump.

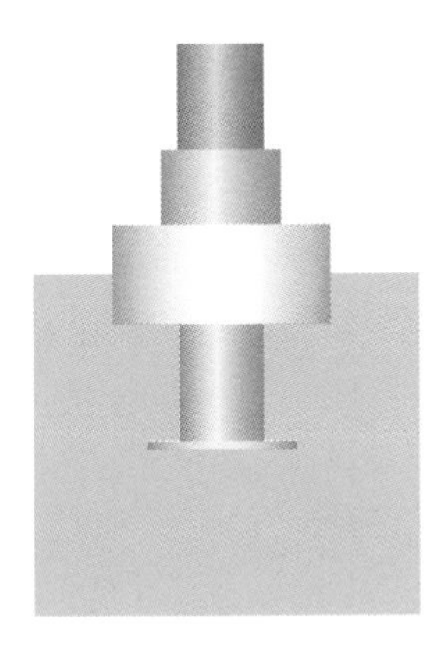

As the water level rises, the valve switches off the pump.

132 Top up evaporation losses with pure water

When you top up evaporation losses, *always* use R.O. water, adding it into a region of strong flow and being careful not to pour it directly onto any corals. Only pure water is lost by evaporation; if you were to use saltwater to top up, your salinity would gradually rise to the point where everything in your reef would die.

133 Monitoring water conditions in a reef aquarium

Once a reef tank is up and running and you have installed a selection of corals, the most important parameters to monitor are, firstly, calcium and carbonates—alkalinity in "German" degrees of carbonate hardness (dKh) or milliequivalents per gallon or liter (meq/gal. or meq/L). These will give you an indication of the quantities of calcium and carbonates being taken up by the corals, giving you a guide to the amount of supplementation necessary. Secondly, you should monitor phosphate and nitrate levels. Consider them as fertilizer for pest algae and keep them to a minimum. Phosphate, in particular, also has the effect of limiting calcification in corals.

134 What water circulation do I need in my tank?

Water movement is of primary importance in the reef tank. Although it is possible to quantify water movement in terms of tank turnover per hour (12 to 20 times tank volume per hour would be a good starting point), it is not just the amount of movement that matters. At the minimum, the pattern of water movement should be chaotic or random. Ideally, however, you should aim to provide some form of surge effect, with water traveling through the aquarium in one direction and, after a given period of time, returning in the opposite direction.

135 Changes in water flow patterns are important

Surfaces in the aquarium, rocks and corals, all have what is termed a "boundary layer." This is a thin region of water next to the surface, which is not really influenced by water flowing by at higher velocity. Consequently, biofilms made up of algae or bacteria can form on some surfaces. Water flow will have little influence on these films unless you aim a powerhead at them at close range. However, by employing some form of alternating water flow or surge, these films can be prevented from forming in the first place or be broken down by the to-and-fro action of the water pulling at the film.

Tip 135 *Introducing water flow into the aquarium helps prevent the formation of biofilms.*

Better Fish Buying

136 There are good reasons for buying captive-bred fish

With the number of captive-bred species steadily increasing, it is a responsible act to buy these fish, rather than continue indirectly to encourage the importation of wild-caught species, thus possibly depleting stocks in the wild. Furthermore, captive-bred fish will be acclimatized to artificial salt mixes and to prepared foods, and are more likely to be disease-free than wild-caught fish. This will greatly enhance your chances of success with these animals.

Tip 136 *Many clownfish are captive-bred.*

137 Why are some fish more expensive than others?

Price is not everything. Several factors influence the final retail price of a fish specimen, from the moment of capture until it is offered for sale in a dealer's tank. Bear in mind that specimens imported through a retailer will be more expensive than those imported directly through a wholesaler. However, the former scenario can mean that the retailer has been able to select his or her own specimens for quality and vitality, and therefore a higher price might be justified.

138 Two clownfish with the same common name

Amphiprion ocellaris and *A. percula* are both sometimes sold as the common clown, so confusion reigns (*A. ocellaris* is also called western clown and *A. percula* is often called percula clown). Some authorities consider the two species as a single, related species, while others classify them separately. Both will readily associate with *Heteractis magnifica* or carpet anemones (*Stichodactyla* spp.) As far as the hobbyist is concerned, does the identity matter? Many authorities would say yes it does, because there is a commonly held belief that *A. ocellaris* is by far the hardier of the two.

139 Be sure to collect up-to-date information

Gathering different opinions is often useful, but beware of old information. Many of the books written on saltwater subjects do not take into account the advances made in the hobby. Improvements in factors, such as filtration, diet, specimen capture and transport, have all assisted in making many species that were once considered to be difficult or even "impossible" to keep no longer so. Consult a few local dealers for reassurance if you are uncertain about the long-term maintenance of a particular species.

140 Ordering specific fish has many advantages

Increasingly, fish collectors are adopting an "order then catch" strategy, whereby dealers place an order from a list of fish species and the selected fish are then captured by the collector. This means that species are not caught and retained by the collector, who must then try to sell them on. Such a strategy reduces the number of individual fish caught and can help in the prevention of overfishing. It should also mean that the fish in the dealer's tanks have a greater vitality, as they have not been languishing in holding tanks for long periods before being exported.

141 Saltwater fish can become your long-lived companions

Many popular saltwater aquarium fish are extremely long-lived. Clownfish can live to be well over 12 years of age, and some species of true angelfish in captivity are more than 20 years old and still going strong. Thus, the choices you make when you first stock your aquarium are very important, as you may be sharing your living room with these fish for many years to come!

Learn the scientific names of saltwater aquarium fish

When requesting information about a particular species from a dealer or fellow aquarist, it is useful to use the scientific (Latin) name to avoid any potential confusion. Certain species will have several common names but only one scientific name. It may seem like a bit of a mouthful but it could prevent the acquisition of the wrong fish.

143 Fish may vary from one part of the world to another

In the case of certain fish, two examples of the same species may show different traits. One will be easy to keep, while the other supposedly identical fish fades away. The beautiful flame angelfish *(Centropyge loriculus)* is a case in point. Some flame angelfish in dealers' tanks have more vibrant coloration and generally look better when compared to other specimens. Closer enquiry usually reveals that the better-looking specimens are from Hawaii, while the less impressive ones often originate in the Philippines. However, there has been a suggestion that the difference in the vibrancy of color of the fish is not just about a geographical shift. Fish from Hawaii have generally experienced improved catching and handling techniques, along with better holding and transportation conditions. You may have to be particularly discerning when choosing your specimen.

Tip 143 *The flame angelfish* (Centropyge loriculus) *lives up to its common name.*

Better Fish Buying

144 Give new fish a chance to settle in before buying them

You should allow for the fact that fish offered for sale may well be new arrivals and, as such, require a little time to settle into their new surroundings. They may exhibit travel fatigue and signs of needing a good meal. While you may be excited at the thought of acquiring a new fish, leave it with the dealer to regain its composure (and fuller figure) and return a few days later to reassess your choice.

145 Buy your fish during quiet times at the dealer

Once you have established which dealers you intend to buy from, try to visit them during their quietest periods. You should find that they will have more time for you on a one-to-one basis and are able to provide more complete answers to your questions. However good the dealer is, remember that their livelihood depends upon selling livestock at a profit, not dispensing advice for free, although customer satisfaction will always be high on the list of priorities for successful retailers.

146 Show your wish list to a knowledgeable dealer

Compiling a list of fish species you would like to obtain for your aquarium can be done well in advance of first stocking. Showing this to knowledgeable dealers and soliciting their advice can help you to minimize aggression and territoriality issues before they arise. You can also devise a stocking order, starting with the hardier species and finishing with the most delicate or territorial species. A good dealer may also be able to suggest alternatives that might be more appealing.

147 How do I know if I am buying a healthy fish?

One of the best pointers to look for is the fish's breathing rate. Fish pump water over their gills in order to extract the oxygen. By looking at the rate with which their gill covers, or opercula, move up and down, you can obtain a reliable impression of their state of health. Unhappy or sick fish may appear to be panting, with a rapid breathing rate, and you should avoid buying them.

Tip 145 *Look carefully at the fish you intend to buy and seek advice if you are unsure about any of them.*

Fins on a saltwater fish

Dorsal fin
The main "keel" to prevent the fish rolling during swimming.

Pectoral fin (paired)
Many fish use these fins to help steer and change position.

Caudal fin (tail)
Most fish swim by sweeping the back part of the body from side to side. The tail fin helps to convert that movement into forward thrust.

Pelvic fin (paired)
These also help the fish to control its position.

Anal fin
This single fin acts as a stabilizer.

148 Not all saltwater fish have pelvic fins

Take your time looking at the fish in the dealer's tanks before making any buying decisions. Do not be alarmed, however, if some fish appear to be lacking pelvic fins. Some triggerfish, filefish and boxfish do not have them.

149 Many saltwater fish swim with folded fins

Freshwater tropical fish frequently swim with folded fins when suffering from disease. Do not be put off from buying a saltwater fish with folded fins—many saltwater fish swim this way when quite healthy.

Tip 149 *Blue damselfish* (Pomacentrus pavo) *are among many saltwater fish that swim with folded fins.*

150 Do not disregard a saltwater fish with damaged fins

Saltwater fish have remarkable powers of regeneration, and specimens that show evidence of slightly split or damaged fins should not necessarily be shunned. Provided the individual concerned is feeding well and shows no signs of disease, you can be confident that the fins will repair in a matter of days. If in doubt, ask if the fish can be reserved for a couple of days before you take it home.

151 Don't ignore a fish with less than vibrant color

Fish that have been resident in a dealer's aquarium for a long time sometimes experience slight color loss. This is probably because the retailer's tank is a little less comfortable than a home aquarium. However, do not feel you should avoid buying these fish. If they have survived in such an aquarium for a long period they should be well adjusted to captivity and feeding readily. Transfer to a home aquarium invariably results in the return of full color after only a short time.

Better Fish Buying

152 Make sure potential fish purchases are feeding

Most dealers won't mind if you ask to see a fish actually eating before you buy it. If necessary, make a return trip to the store to confirm this. Unwilling feeders may not change their ways once you get them home.

153 Don't let your heart rule your head

Never allow a fish's beauty or rarity to cloud your judgment. Often, the desire to own a particularly attractive specimen can overrule your usual checklists for purchases. If there is a question mark over the condition of a fish, it is best to leave it alone and visit the retailer at a later date to check on its progress. The fish concerned may have been sold in the meantime, but that is the risk you must take. Better that than to jeopardize the well-being of the entire aquarium through a rash purchase.

154 Leave room in the tank for future acquisitions

Never be in a rush to fill your aquarium with fish. Many saltwater species are seasonal and some have very short collection periods. There is little worse than falling in love with a species of fish previously unknown to you but that you cannot have because there is no more room in your aquarium.

155 How many fish will my tank support?

As a guide to stocking levels in a reef system, allow 1 inch (2.5 cm) of fish per 6 gallons (23 L) of water. In a fish-only system, allow 1 inch (2.5 cm) of fish per 3 gallons (11.5 liters). Stick to these levels for the first year of a tank's life. There can be no hard and fast rules as no two aquariums are ever identical, but by being conservative during the first year while the tank is maturing and stabilizing, your chances of success are greatly enhanced.

Tip 156 *Do not be tempted to overstock the saltwater aquarium with fish.*

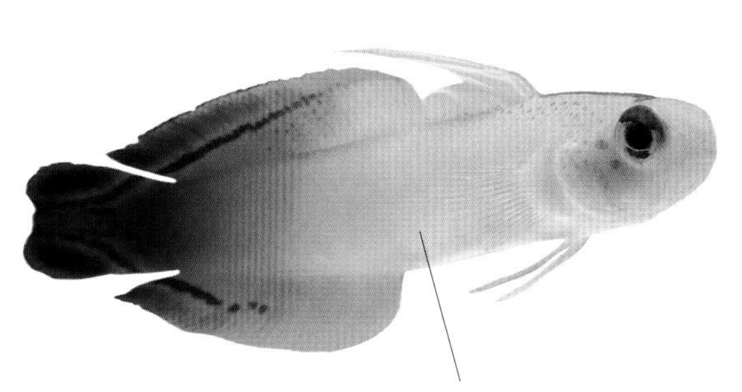

Tip 158 *Fire goby* (Nemateleotris magnifica).

156 Don't allow yourself to be carried away

It is easy for newcomers to the saltwater hobby to fall into the trap of overstocking the aquarium in their enthusiasm to achieve a stunning collection of fish. Be patient. Stringent stocking levels may seem harsh, but they are essential. Do not aspire to an instant community; instead, build up stocking levels slowly over a period of months, not weeks, until the tank is at its peak maturity. Allow for growth and the final size of the species. You will see differing stocking guidelines published; use them sensibly. For example, a 6-inch (15 cm) porcupinefish (*Diodon* spp.) is going to produce more waste and will be a far greater burden on the tank than three 2-inch (5 cm) damselfish.

157 Species that contribute to the success of a reef aquarium

In a reef aquarium it is worth introducing species that will undertake useful roles, and therefore put less overall pressure on the system, rather than fish that simply take and give nothing back. A good example is the yellow sailfin tang *(Zebrasoma flavescens)*, which grazes on macro-algae. It is very useful in the prevention of filamentous forms of algae. Introducing this species with a bristletooth tang (*Ctenochaetus* spp.) is considered by many authorities to be an excellent and highly beneficial combination of fish.

158 Start a reef aquarium with easy, reef-friendly fish

When starting out with your first reef aquarium, choose species of a size proportionate to the size of your tank. Stick to easy-to-keep, small, compatible, reef-friendly fish, at least for the first year of your reef aquarium's life. Best choices would be small planktivores and small herbivores. Try to avoid any animal that might look on a fellow tankmate as a possible snack.

159 Go for groups in the reef aquarium

Whenever possible, try to keep fish in natural groupings. Given the relatively small scale of a captive reef, this is not always possible, so choose fish that can be kept in natural groupings in your size of reef aquarium, and you will be rewarded with happier, less stressed creatures.

Tip 159 *A group of regal tangs* (Paracanthurus hepatus) *need a large aquarium.*

160 A host anemone is not always required

Not all clownfish require an anemone in the tank. Independent species include cinnamon *(Amphiprion melanopus)*, tomato *(A. frenatus)* and maroon *(Premnas biaculeatus)*. Dependency on anemones may be receding as more and more clownfish are bred in captivity. (See also Tip 328.)

Tip 160 *The tomato clownfish* (Amphiprion frenatus) *does not need an anemone.*

161 Other partnerships are possible in the aquarium

Saltwater fishkeepers are quite likely to keep clownfish (*Amphiprion* spp.) and a sea anemone together, but this is not the only symbiotic combination possible in the aquarium. Shrimp gobies (*Amblyeleotris* and *Cryptocentrus* spp.) often share a close relationship with pistol shrimp (*Alpheus* spp.), both sharing a common burrow in the substrate. Keep this in mind when you are stocking your aquarium.

162 Find out how large your chosen fish will grow

Remember that in nature there is a continuous food chain in operation. Big fish with big mouths will eventually eat smaller fish, so check out what size your chosen fish will attain before buying it. Without exception, fish at the dealer's are always juveniles and most will have a great deal of growing to do.

163 Some species aren't so appealing when fully grown

The panther, or polkadot, grouper *(Cromileptis altivelis)* is a very appealing fish, especially when young, but do not be fooled by the small specimens for sale at a little over 2 inches (5 cm). The polkadot grows rapidly in captivity and can easily reach 9 inches (23 cm) (up to 28 inches or 70 cm in the wild), so it needs to be housed in a tank capable of coping with its large size. Also be aware that, as a predator, the polkadot grouper will become a threat to other tankmates.

Tip 163 *Polkadot grouper* (Cromileptis altivelis)

164 Some species will outgrow the aquarium

Some fish are just too big to be kept in captivity, so make sure you know exactly what species you are buying. Do not buy a fish sold by a common name if the dealer has no idea as to its true identity. Totally unsuitable potential monsters that you might see offered for sale include certain wrasse, groupers and moray eels.

165 Tiny boxfish are unlikely to survive to maturity

The very attractive and appealing boxfish, in particular the yellow boxfish (*Ostracion cubicus* or *O. tuberculatum*) are often offered for sale as very tiny juveniles, measuring less than an inch (2.5 cm). Bearing in mind that this species can reach 18 inches (45 cm) in its natural environment, you begin to realize just how small these specimens are. Unfortunately, many of them suffer a slow demise once in captivity. Since providing an adequate diet and nurturing such a tiny fish through its vital juvenile development stages in captivity is so challenging, aquarists should not buy these difficult fish when they are at such an immature stage.

166 Make a considered choice when buying damselfish

Try to minimize the potential for aggression in pugnacious fish at the point of purchase. For example, when considering a group of damselfish, do not buy the one that is terrorizing the rest of the group, nor the one that seems to be an outsider and subject to constant bullying. Avoid both the largest and the smallest specimens. If you intend to house a group, buy all the fish at the same time and ensure they are of a similar size. It will be difficult to add further damselfish to an established group unless they are very small juveniles. Even when adding other, totally unrelated fish, a large established damselfish will almost certainly react aggressively toward the newcomer if it is of a similar size or color.

167 Some saltwater fish are more aggressive than others

Some saltwater fish, such as triggerfish, are always going to cause you headaches on account of their aggression toward virtually any tankmate. It is not always easy to predict which species are going to cause problems. For example, undulate triggerfish *(Balistapus undulatus)* and queen triggerfish *(B. vetula)* are undoubtedly very vicious individuals. The striking clown triggerfish *(Balistoides conspicillum)* is very popular and usually aggressive, but if purchased as a small specimen may tolerate other fish. The Picasso triggerfish *(Rhinecanthus aculeatus)* is another species with a varied reputation. Triggerfish from the *Sufflamen, Xanthichthys* and *Melichthys* genera may be more amiable. The important thing is to be sure of the identity and reputation of your intended purchase, and to seek advice before committing yourself.

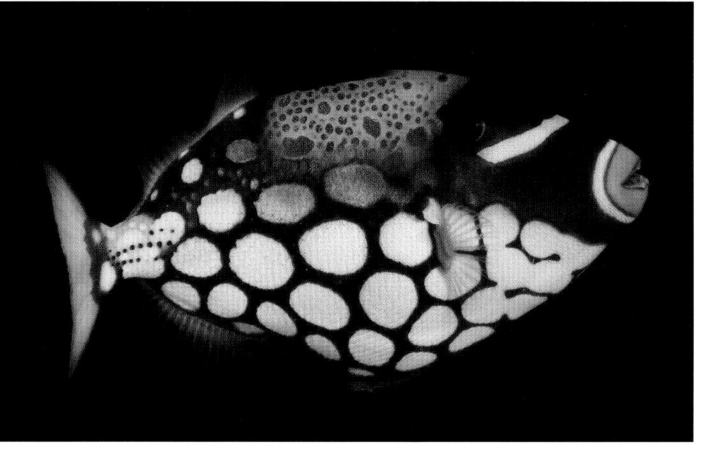

Tip 167 *Clown triggerfish* (Balistoides conspicillum)

168 Acquiring a known pair of wreckfish is not a problem

Sexing fish can be difficult, which makes acquiring pairs a problem. However, the fish themselves can often solve the problem for you naturally. For example, if you buy a small number of wreckfish (*Pseudanthias* spp.) they will conveniently produce a male (through sex-reversal) from a number of females, as and when required. (See also Tip 436.)

169 Some species of fish are not what they seem

Do not be misled by lookalike species. Some are benign, such as the mimic surgeonfish *(Acanthurus pyroferus)*, which is very similar to the young lemonpeel angelfish *(Centropyge flavissimus)* and the half-black angelfish *(C. vroliki)*. Others are not so safe. Do not confuse the cleaner wrasse *(Labroides dimidiatus)* with the sabertooth blenny *(Aspidontus taeniatus)*, as the latter species takes advantage of its cleanerfish-like markings to approach unsuspecting fish, from which it takes a chunk of flesh instead of an irritating parasite.

170 Choose the most suitable wrasse for your system

There are over 60 genera and 600 species of wrasse, so even if only a few are suitable for the hobby it still leaves a great deal of choice for the hobbyist. Species can be as small as 2 inches (5 cm) or up to 60 inches (150 cm) when fully grown in the wild. Identifying small wrasse, especially juveniles, while problematic, is essential for success. Many mature wrasse become too large and destructive for captivity, and can also become increasingly aggressive. Research your intended purchases with care.

171 Many angelfish change pattern as they mature

Do not rely on a group of young angelfish growing up to become identical adults. Young angelfish are often blue with white markings and change to completely different color patterns as they mature.

Tip 171 *Juvenile specimens of the emperor angelfish* (Pomacanthus imperator) *have white markings on a blue background (above). The adult emperor has a distinctive patterns of blue lines on a yellow body and a yellow caudal fin (left).*

Tip 172 *Cleaner wrasse* (Labroides dimidiatus)

172 Is the cleaner wrasse a suitable aquarium subject?

It depends on how you look at things. The cleaner wrasse *(Labroides dimidiatus)* performs an important service in the wild by pecking off parasites from the skin and gills of infected fish that visit its area. It could be argued that, in the pristine conditions of the home aquarium, fish are less likely to be infected, so the cleaner wrasse will be denied its natural diet. Some authorities consider that cleaner wrasse should not be collected from the wild because their natural clientele are showing signs of poor grooming. However, reports seem to indicate that the cleaner wrasse does adapt to aquarium life fairly easily.

173 A substitute for the cleaner wrasse in the aquarium

If you want a cleanerfish, why not opt for the tank-raised neon goby *(Elacatinus oceanops)* or even a cleaner shrimp *(Lysmata amboinensis)*. The goby will not survive to a great age—perhaps only a couple of years—but this is because it is a short-lived fish, not because it fares badly in captivity. On the contrary, it seems to do very well, even spawning quite readily.

174 Tempting, but not intended for life in captivity

The blue ribbon eel *(Rhinomuraena quaesita)* is very striking, with a number of color variations. Depending on sex and maturity, the slender, flowing and graceful body varies from bright blue to almost black, with a contrasting yellow dorsal fin running the entire length of the eel. Such a beautiful specimen is bound to attract attention, but unfortunately it is virtually impossible to keep in captivity. It will often refuse to eat and slowly starves to death. Even if you manage to wean it onto live food, it simply does not adapt to captivity. Do not be tempted into trying to keep this fish.

Tip 175 *The cowfish* (Lactoria cornuta)

175 Cowfish can emit a toxin lethal to all fish in the tank

In order to defend itself, the cowfish *(Lactoria cornuta)*, and all its trunkfish and boxfish relatives, can release a toxin that is lethal in minute quantities to all fish, including the cowfish itself. In the wild, the fish would emit its toxin and then swim away from the danger, but this is not possible in the closed confines of an aquarium. Here, all the fish will die very quickly—far too quickly for you to act to save them. If you keep one of these fish it must not be upset. Aim to avoid any stress, including unsettling tankmates, in hopes of avoiding the deadly outcome of a toxin release.

Better Fish Buying

Tip 178 *Float the transport bag in the tank.*

176 Transport cowfish with minimal stress

Cowfish *(Lactoria cornuta)* are liable to produce their lethal toxin if stressed during transportation, so make sure their journey home is as uneventful as you can make it. Transport the fish separately (and singly) as any toxin they produce will kill them and everything else in the bag.

177 Match the water conditions when transferring fish

To minimize stress when transferring fish from your quarantine tank into the display aquarium, make sure that the water conditions in the quarantine tank match those in the main aquarium. Ideally, transfer fish in subdued light.

178 Acclimatizing new fish after a long journey home

With their low tolerance to changes in conditions, acclimatizing new fish is tricky. Following, say, a long journey home, the conditions in the transportation bag may have deteriorated so much that it is best to transfer the fish immediately into better quality water. Most acclimatizing methods advocate against adding too much of the transportation water to the quarantine tank for fear of introducing pollution or disease.

179 Check water salinity levels during acclimatization

When going through a gradual acclimatization process of water exchange, remember to check that the salinity in the bag matches that of the aquarium before introducing the animal, be it fish or invertebrate.

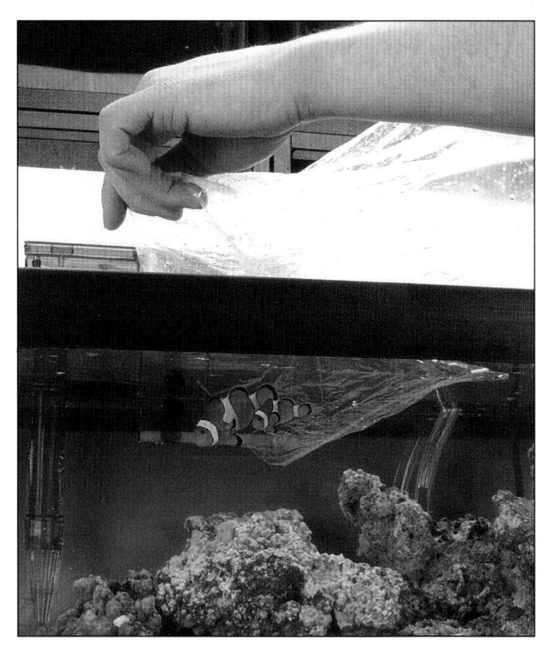

Tip 178 *Release the fish when water conditions are equalized.*

180 Cover the tank to prevent new fish jumping out

If your aquarium is normally left open, it is a good idea to keep it covered for the first few nights following the introduction of new fish. Until they have settled in, many fish have a tendency to jump at night if they are spooked by other tank inhabitants. Once they have found a comfortable niche where they feel safe, they will settle down.

181 Exercise a little territorial trickery when adding new fish

When stocking the aquarium, bear in mind the nature of the fish you introduce. Some damselfish, for example, become very territorial and, once installed, are likely to resent heartily any newcomers from that moment on. To avoid territorial squabbles when adding new fish, try a little psychology. Rearrange some of the rockwork slightly just beforehand; this forces the resident fish to stake out their territories again, rather than deliberately pick on any newcomers.

182 Coping with bullying in the aquarium

Sometimes a bullying problem occurs after you add a new fish to a tank. The new fish can be either the aggressor or the victim. If the situation does not improve, you may have to intervene to tame the bully. One possible, but not foolproof, method is to maintain the tank in complete darkness for 48 hours by draping it with a heavy blanket. Theoretically, during this period of enforced slumber, the newcomer will be readily accepted and/or the bully will have calmed down. Another strategy is to place a clear plastic divider in the tank for up to two weeks to separate the dominant bully from the rest of the tank inhabitants. A further step is to remove the aggressor to another tank for a week or two to allow the new fish to settle into the display aquarium. On its return to the main tank, the bully will be disoriented and will have calmed down. However, the stress placed on the bully could permanently disrupt its feeding pattern. Ultimately you may have to return the bully to the dealer, or if the victim of the bullying is a single fish, return the victim instead.

Tip 181 *Rearranging the decor can prevent territorial squabbles.*

183 Choose fish species after careful consideration

Remember, it can be very difficult to extract unwanted fish from a heavily aquascaped tank; you may even have to strip down the tank to remove them. So do think carefully before introducing fish and make sure that they are a species that you really want to keep.

184 You may need to use a commercial fish trap

Rather than dismantle your aquarium when it becomes necessary to remove a fish that is too aggressive or has outgrown your system, it is well worth acquiring one of the commercial fish traps that are designed to facilitate the capture of problem fish with minimum stress to the target fish and its tankmates. Most designs use food as bait and a simple, manually operated trap-door to catch the fish. They will justify their purchase price a number of times over during an aquarium's lifespan.

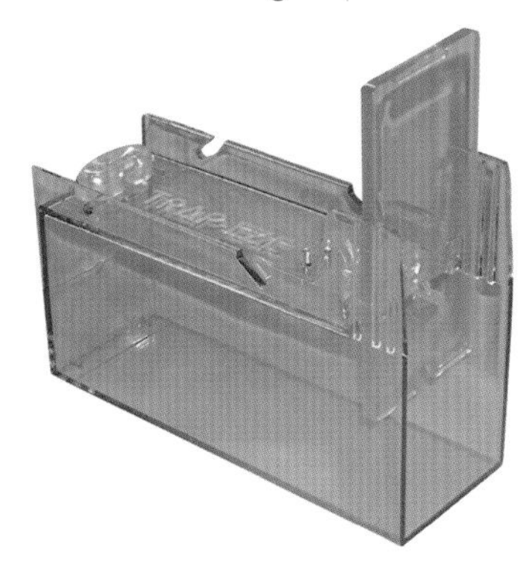

Tip 184 *A fish trap is safe and easy to use.*

185 Some fish grow too large for the home aquarium

It is doubtful whether batfish are suited to being kept by hobbyists unless they can be housed in an aquarium of 200 gallons (750 L) or more. Along with many other large fish available in the hobby, batfish can constitute a problem for the public aquariums to which they are invariably offered, having outgrown the hobbyist's facilities.

186 Capturing fish with spines or scalpels

Some fish, including squirrelfish and angelfish, have sharp spines that can cause major problems if you need to capture them. The same applies to surgeonfish with their scalpels. The spines or scalpels can become entangled in a net, leading to extreme stress for the fish (and hobbyist). If you have to catch such fish, use a small clear plastic container or a net of exceptional quality.

187 What to do if you are stung by a lionfish

Take care during tank maintenance, because if you corner a lionfish, also called turkeyfish, in its own territory you stand a chance of being stung—an excruciating experience. If this should happen, the crucial first aid treatment is to flush the area with water as hot as you can bear (but without adding to the problem by scalding) as soon as possible after the stinging incident. It is a good idea to seek medical help as the wound will need cleaning, and antibiotics may be prescribed to prevent any infection developing in the wound.

Tip 187 Pterois volitans, *a lionfish that can deliver a painful sting.*

188 Seek advice before choosing a large angelfish

To keep large angelfish successfully, you need a fair amount of experience; they can never be classed as easy fish. Some species are somewhat less challenging than others, but always seek the advice of fellow aquarists and reputable, knowledgeable dealers. The following species are generally considered to be the hardiest, although "hardy" is a relative term: French angelfish *(Pomacanthus paru)*, Koran angelfish *(P. semicirculatus)*, gray angelfish *(P. arcuatus)*, queen angelfish *(Holocanthus ciliaris)* and blue angelfish *(H. bermudensis)*. Despite their visual appeal, angelfish present a challenge even to very experienced fishkeepers.

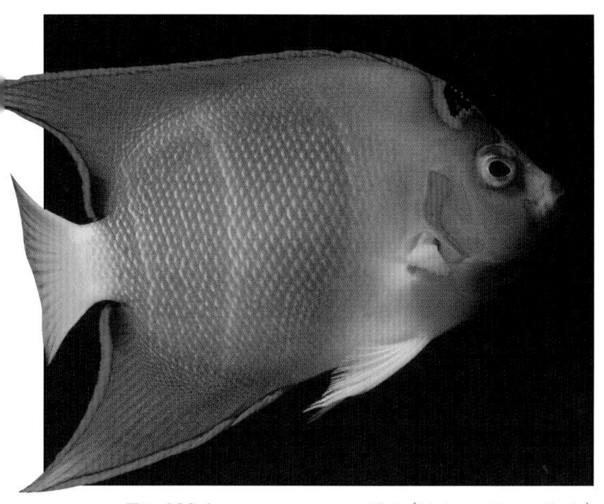

Tip 188 *A young queen angelfish* (Holocanthus ciliaris).

189 Large angelfish need a sufficiently large aquarium

If you are going to keep any angelfish, make sure you provide them with a sufficiently large aquarium. These fish can reach sizes of 14–20 inches (35–20 cm) depending on the species, so an appropriate volume would be in the region of 132–185 gallons (500–700 liters), again depending on species.

Tip 190 *Majestic angelfish* (Pomacanthus navarchus)

190 Can I keep a large angelfish in a reef aquarium?

Not all true angelfish (family Pomacanthidae) are unsuitable for reef aquariums. Provided you are vigilant and prepared to forgive a little coral-nipping should it occur, some species are reasonably trustworthy. Favorites among aquarists include the majestic angelfish *(Pomacanthus navarchus)*, regal angelfish *(Pygoplites diacanthus)* and emperor angelfish *(Pomacanthus imperator)*. Be aware that the emperor is more commonly stocked as a small juvenile and may become a nuisance when it changes into its adult coloration.

191 Some less familiar angelfish make good marine subjects

Angelfish of the genus *Genicanthus* are less well known than many of their more showy relatives, but they make excellent additions to most saltwater aquariums. All the species contained in this group feed on zooplankton, meaning that they swim in open water, readily accept frozen or particulate foods and are rarely a problem with sessile invertebrates. There are also significant color and pattern differences between males and females, allowing pairs to be easily identified and kept in larger aquariums.

192 Dwarf angelfish are small but perfectly formed

Dwarf angels are small angelfish from the genus *Centropyge*, mainly ranging from 2.5–6 inches (6–15 cm) in length. They belong to the same family (Pomacanthidae) as their larger cousins. They are not only attractive, colorful and active, but are hardier when kept in captivity. They can make a great choice of specimen fish for the smaller tank, i.e., in tanks too small to accommodate a tang as a specimen fish.

193 Will dwarf angelfish damage corals?

Some dwarf angels have a tendency to nibble at corals. This activity may be confined to browsing on sugar- or bacteria-rich mucus or it may result in damage to certain corals. The potential for damage varies from fish to fish and from species to species, but nevertheless, many aquarists love them in a reef tank for their beauty. The smallest of these, the cherub angelfish *(Centropyge argi)* can probably be considered harmless due to its size. Another very popular species is the flame angelfish *(Centropyge loricula)*, a stunningly bright red fish. Think carefully before taking on any of these fish. They are beginner-friendly but may affect your reef in ways you might find undesirable.

194 You can mix species of dwarf angelfish

Contrary to what has been written in many saltwater fish guides, dwarf angelfish of the genus *Centropyge* can be mixed, albeit with a degree of caution. Of them all, the coral beauty *(C. bispinosa)* is the most relaxed and therefore least territorial. Thus, if you are considering including this species plus another couple of dwarf angels, put the coral beauty in first to minimize any conflict.

195 Dwarf angels and tangs browse on saltwater algae

Species that spend a great deal of their natural lives browsing on algae, such as dwarf angelfish and tangs, will not only benefit from the presence of live rock in their aquarium, but also from dried marine algae held in a lettuce clip. They can pick at this food source over a much longer period than they would be able to feed at frozen or particulate foodstuffs. As well as simulating the natural way of feeding among these fish, it can also assist in reducing any tendency to nip at sessile invertebrates. (See also Tip 348.)

Tip 194 *Coral beauty* (Centropyge bispinosa)

196 Herbivorous blennies help to keep algae under control

There are many species of peaceful, small blennies to choose from. They can be roughly divided into herbivores and planktivores. Herbivores include the bicolor blenny (*Ecsenius bicolor*) and jeweled rockskipper, also called algae blenny (*Salarius fasciatus*). Both species keep algae under control in the reef. Do not mix species of herbivorous blennies in the smaller aquarium, as this can lead to fighting. Planktivores can be typified by the Midas blenny *(Ecsenius midas)*, an endearing, colorful species.

Tip 196 *Jeweled rockskipper* (Salarius fasciatus)

197 The bicolor blenny has much to offer

Ecsenius bicolor may lack the beauty, coloration and nonstop activity of many fish, but its quirky nature, hardy disposition and eagerness to feed more than compensate. In addition, it poses few disease or compatibility problems and is also cheap to buy—a not inconsiderable advantage. The bicolor blenny forages outside its retreat and then reverses back into its home to survey its surroundings from a secure vantage point. Fish such as these are often a better proposition until you have gained a little experience in saltwater fishkeeping.

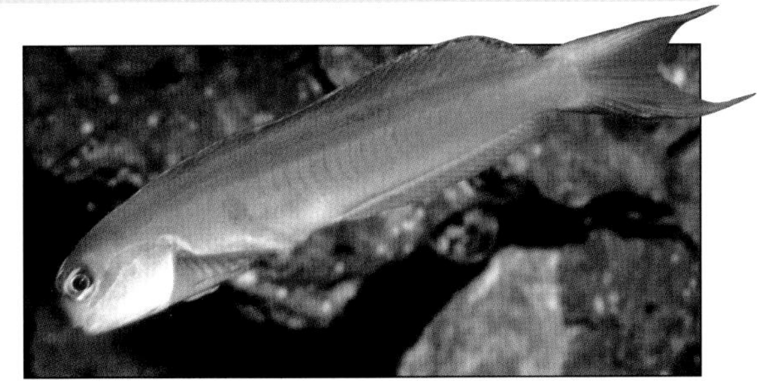

Tip 198 *The Midas blenny* (Ecsenius midas)

198 The Midas blenny does well with large tankmates

The Midas blenny *(Ecsenius midas)* swims actively in the water column, in stark contrast to many other members of the genus. It is often found in association with schooling wreckfish (*Pseudanthias* spp.), which it closely resembles. It is a superb choice for an aquarium containing larger fish species, although it can bully similar-sized planktivores.

199 The Red Sea mimic blenny is worth seeking out

The Red Sea mimic blenny *(Ecsenius gravieri)* is a beautiful species that mimics the blackline fang blenny *(Meiacanthus nigrolineatus)*, which possesses venomous fangs. However, the mimic is as harmless as its close relative the bicolor blenny and is hardy, peaceful and full of character. This species is an absolute gem, provided that you can find one.

Tip 199 *Red Sea mimic blenny* (Ecsenius gravieri)

200 Does an "easy" butterflyfish exist?

Most people would agree that "easy butterflyfish" is a contradiction in terms. However, a few species are generally regarded as easier than others, and when you have gained some experience, these are the ones to consider first. They are threadfin butterflyfish *(Chaetodon auriga)*, vagabond butterflyfish *(C. vagabundus)*, Klein's butterflyfish *(C. kleini)*, pearlscale butterflyfish *(C. xanthurus)*, black-backed butterflyfish *(C. melanotus)*, millet-seed butterflyfish *(C. miliaris)* and raccoon butterflyfish *(C. lunula)*. One important factor to consider is the quiet nature of butterflyfish; do not mix them with boisterous tankmates.

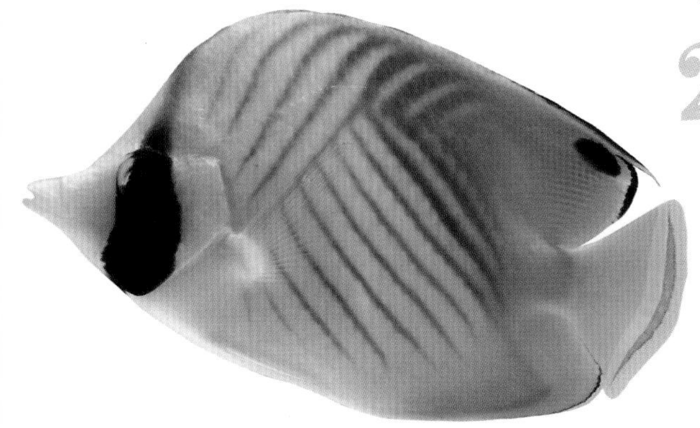

Tip 200 *Threadfin butterflyfish* (Chaetodon auriga)

201 Small fish can be aggressive too

Do not overlook smaller fish as a source of aggression. Most hobbyists are aware of the potential bullying a damselfish can dish out, but far fewer realize that members of the Pseudochromidae family are persistent fin-nippers. An old staple of the hobby, false gramma, or royal, dottyback *(Pseudochromis paccagnellae)* is losing popularity with hobbyists because of this unsavory streak.

Tip 202 *Copperband butterflyfish* (Chelmon rostratus)

202 Copperband butterflyfish need help with feeding

The copperband butterflyfish *(Chelmon rostratus)* has an ill-deserved reputation as a poor feeder in captivity. The long snout is an adaptation for browsing small saltwater organisms among the nooks and crannies of the reef, and these fish can be outcompeted for food in open water. One possible feeding strategy is to pack a food such as frozen mysis shrimp into an empty shell and position it in an area of the tank frequented by the fish, preferably away from where other fish are feeding. Only keep these fish in a reef aquarium with live rock where they can find tasty morsels between main feeds.

203 The royal gramma is a brilliantly colored fish

The royal gramma *(Gramma loreto)* is a very striking fish with its half-purple, half-yellow coloring. Keep one large or medium fish with one or two smaller ones. Do not confuse it with the similarly colored false gramma or royal dottyback *(Pseudochromis paccagnellae)*, an aggressive species from the western Pacific.

204 Some dottybacks are more peaceful than others

Specimens of dottyback (*Pseudochromis* spp.) that are endemic to the Red Sea and Arabian Gulf are often more peaceful than their Indo-Pacific counterparts. Examples of such species include the orchid (*Pseudochromis fridmani*), sunrise or blue (*P. flavivertex*), orange or neon (*P. aldabraensis*) and striped dottyback (*P. sankeyi*). The Red Sea species are often available as captive-bred individuals.

Tip 204 *Sunrise, or blue, dottyback* (Pseudochromis flavivertex)

205 Damselfish are a hardy choice for the beginner

Damselfish are exceptionally hardy and a good choice for newcomers to the saltwater hobby. They are also inexpensive, small, colorful and constantly active. However, they do display overt aggressiveness toward each other and to other, unrelated species.

206 Two more species of peaceful damselfish

Many damsels can be more trouble than they are worth in a mixed reef, but if you are happy to stick to just one species of fish in your tank they can be very rewarding. They are a good alternative to clowns, but do not think of them as something to keep for a while and then remove before going on to something else. Recommended species would be yellowtail blue damselfish (*Chrysiptera cyanae*) or Allen's damselfish (*Pomacentrus alleni*). Both are among the most peaceful of the damselfish and can be kept quite easily as pairs or in groups in a well aquascaped tank.

207 Blue-green chromis tolerates its own kind

Like other members of its genus, the blue-green chromis (*Chromis viridis*) is quite tolerant of other individuals of the same species. As a result, it is an ideal species to introduce if you want to have a school of fish. In addition to their hardy disposition, they also serve a useful role in encouraging new additions to the aquarium to swim into open water. They also have the added incentive of being one of the most reasonably priced of all saltwater fish.

Tip 207 *A school of blue-green chromis* (Chromis viridis).

Better Saltwater Fish

208 Wait before introducing specialized feeders

Certain fish require foodstuffs such as copepods, amphipods and other small crustaceans that occur naturally in reef or live-rock-based aquariums. However, in the first months of the aquarium's existence, the populations of these animals are unlikely to be stable, and the predation pressure exerted by a hungry fish may cause a collapse in numbers from which these tiny creatures cannot recover. For this reason it is best to add species such as dragonets and mandarinfish (*Synchiropus* spp.) after the aquarium has been established for at least six months, so that crustacean populations have had time to stabilize. Also keep in mind that many people make the mistake of adding too many species that compete for a single resource. A common example is adding a pair of mandarins to an aquarium with a pair of scooter dragonets.

209 Fish that may help to control planarian flatworms

Green mandarinfish *(Synchiropus picturatus)*, sixline, or a pajama, wrasse *(Pseudocheilinus hexataenia)* and scooter blennies (*Synchiropus* spp.) are just three fish species that have been quoted as useful in the control of the unsightly planarian flatworm *(Convolutriloba retrogemma)*. However, the efficacy of each species depends greatly on the individual fish concerned. Some work well but others will ignore the flatworms completely. Only introduce any of these species if they were included in your original stock list.

210 Tangs need a tank of reasonable size

Many aquarists are very fond of the yellow tang *(Zebrasoma flavescens)* and, fortunately, it is a fish that does well in the aquarium. An alternative choice would be the brown, or scopas, tang *(Z. scopas)*. These are two of the smallest tangs, but you should nevertheless ensure that they are housed in nothing smaller than a 66–gallon (250 L) tank. As they are herbivores, you should supplement their diet with saltwater greens.

Tip 210 *The yellow tang* (Zebrasoma flavescens) *relishes green food.*

211 Introduce surgeonfish to the aquarium last of all

With the notable exception of the two species of mimic surgeonfish *(Acanthurus pyroferus* and *A. tristis)*, surgeonfish should be the last fish introduced into a saltwater aquarium to avoid any territorial conflicts. Scalpel-like scales located at the base of the tail on the caudal peduncle give this group their common name, and they are capable of inflicting severe wounds.

212 Always buy small specimens of *Acanthurus*

Members of the genus *Acanthurus*, so-called surgeonfish, are notoriously aggressive. If you must have one, then look to species that can be obtained at a small size, in many cases only an inch or so, to reduce any potential problems. Examples include the clown surgeonfish *(A. lineatus)*, blue tang or blue Caribbean *(A. coeruleus)* and Red Sea clown *(A. sohal)*. Provided you allow for their growth and are sensible with subsequent stocking, you can enjoy a surgeonfish without having to endure their territorial aggression.

213 The blue tang has a large appetite and grows quickly

The blue, or regal, tang *(Paracanthurus hepatus)*, is available at sizes much smaller than most tangs or surgeonfish, but grows rapidly and could reach 12 inches (30 cm) or so within a few years. It has a large appetite and can present a threat to sessile invertebrates if you do not offer it enough food throughout the day. Providing dried algae can help to prevent this tendency. However tempting this fish may seem, consider its long-term needs. This is true for many saltwater fish species, but particularly for the blue tang.

214 Think carefully before buying surgeonfish species

Surgeonfish can be prone to contracting an ich, or whitespot, infection *(Cryptocaryon)* shortly after introduction into a new aquarium. If it is a reef aquarium, it can be very difficult to treat. Although there are many ways to treat this disease, none is as effective as copper, which is prohibited for reef aquariums. Consider whether you really must have any of these species before you buy one. Could you really do without the hassle? If you decide that you want one, then quarantine it before adding it to the display aquarium.

215 Bristletooth tangs are hard workers in the aquarium

Members of the genus *Ctenochaetus* are also referred to as bristletooth tangs. Their specialized teeth are adapted to remove algae and detritus from rock surfaces and therefore make excellent additions to most saltwater aquariums, where they earn their keep by doing essential "housework." They are good companions for "proper" tangs, which eat and prevent macro-algae, while the bristletooths consume micro-algae and detritus.

Tip 213 *Blue tang* (Paracanthurus hepatus)

216 Fairy wrasse are beautiful and generally peaceful

Fairy wrasse (*Cirrhilabrus* spp.) make ideal additions to the reef aquarium, as they are generally very peaceful, highly colorful and remain fairly small (most species are less than 5 in./12.5 cm long). There are many species to choose from, including the exquisite (*Cirrhilabrus exquisitus*), Scott's (*C. scottorum*) and the gorgeous but expensive flame wrasse (*C. jordani*). Pairs of each species are sometimes available, and their active swimming in open water makes them doubly attractive to hobbyists. However, they are prone to jumping and are therefore best suited to very peaceful or covered aquariums.

217 The reef aquarium fish with a split personality

It is difficult to offer advice on the pyjama, or sixline, wrasse (*Pseudocheilinus hexataenia*). Some individuals are very hardy and peaceful and, due to their small maximum size, the perfect reef aquarium fish. However, other individuals are fiercely territorial and attack fish much larger than themselves, harassing them constantly. The same is true of the Hawaiian fourline wrasse (*P. tetrataenia*).

218 Possum wrasse do well in a reef aquarium

The beautiful possum wrasse (*Wetmorella nigropinnata*) is becoming increasingly available within the hobby and is one of the best species you can buy for a peaceful reef aquarium. Possum wrasse are easy to sex and pairs can be kept even in a fairly small aquarium (26 gallons/100 L). Achieving a maximum size of about 3 inches (7.5 cm), these small fish need plenty of hiding places and should not be kept with more boisterous species of a similar size.

219 The western clownfish is the classic choice

Clownfish are the perfect reef fish, never straying far from home in the wild. Unless you have a compelling reason for choosing one of the larger or less common species, stick with *Amphiprion ocellaris*. This is the most commonly seen of the clownfish and readily available as captive-bred stock.

Tip 219 *A common species of clownfish,* Amphiprion ocellaris.

220 A natural sleeping bag that offers protection

Parrotfish and some wrasses retire at night, enclosing themselves in "sleeping bags" made from mucus. This may protect them from predators. Another wrasse characteristic is to bury themselves in the substrate.

221 The blue throat triggerfish does well in captivity

One of the easiest triggerfish to maintain in captivity is the blue throat or gilded *(Xanichthys auromarginatus)*. This species feeds on zooplankton in its natural environment and does not usually harm corals. Although you should take care when introducing this species to an aquarium containing small ornamental shrimp, it is quite peaceful and many aquarists keep it in reef aquariums. Pairs can be kept with few problems, and trios (one male and two females) are also acceptable.

222 Tilefish appreciate a peaceful aquarium setup

Tilefish (family Malacanthidae) are represented within the aquarium hobby by two or three attractive species that are all collected from fairly deep water. The skunk tilefish *(Hoplolatilus marcosi)* is a good example. Individuals can be quite nervous and prone to jumping, so a covered aquarium is a must. However, in a peaceful environment they can settle well, accept most foods and swim actively in open water. As with many species that inhabit deeper water, if they have been brought to the surface too quickly they could have difficulties swimming due to gas bubbles in their body tissues. Avoid such specimens.

223 Consider gobies for the smaller aquarium

There are many species of goby especially suitable for the smaller tank. The majority are peaceful and remain small. Recommendation include: Rainford's *(Amblygobius rainford)*, neon *(Elacatinus oceanops)* or its cousin the golden neon *(E. evelynae)*. Neon gobies are cleaner species that groom other fish; neons are bright blue, while golden neons are gold or yellow.

Tip 223 *Rainford's goby* (Amblygobius rainfordi)

224 Fire gobies may be timid to start with, but do settle

The fire goby, or firefish *(Nemateleotris magnifica)*, is a great choice for the peaceful community tank. It is a small planktivore and often timid when introduced. However, once settled it will hover out in the open above its chosen hiding place. Keep singly or in pairs.

Tip 224 *The fire goby or firefish* (Nemateleotris magnifica)

225 The bluestripe pipefish accepts frozen food

The bluestripe pipefish *(Doryramphus excisus)* is probably the best species of pipefish to keep in an aquarium. It is well suited to a peaceful reef, or live-rock-based, fish-only aquarium. You can keep the fish in pairs, and most individuals will feed readily on frozen food. They are certainly a better choice than most other species of pipefish, which can prove very difficult to feed.

Tip 225 *Bluestripe pipefish* (Doryramphus excisus)

226 Stunning, but not really for beginners to keep

The Moorish idol *(Zanclus cornutus)* is a stunning saltwater fish that historically has not done at all well in saltwater aquariums. However, with the advent of live-rock-based systems and improvements in nutrition, the Moorish idol should no longer be considered impossible. The hardiest specimens seem to come from Hawaiian waters, although this may be due to better collection and shipping techniques than to the fish themselves. Even so, this is not a fish for the beginner or even moderately experienced aquarist.

Tip 227 *Fumanchu lionfish* (Dendrochirus biocellatus)

227 Can I buy a reasonably sized lionfish for my tank?

If you would like to keep a lionfish, there are three readily available species that remain quite small. These are the fuzzy dwarf *(Dendrochirus brachypterus)*, Fumanchu *(D. biocellatus)* and clearfin *(Pterois radiata)*. None of these species will grow to more than 8 inches (20 cm) in captivity. Contrast this with the commonly imported turkeyfish, *P. volitans*, which is often simply called lionfish and can reach a length of 20 inches (50 cm).

228 Always keep chalk bass in a group in the aquarium

The chalk bass *(Serranus tortugarum)* is a beautiful, small, non-aggressive, social species of seabass that makes a great beginner's fish. The sexuality of these fish is particularly interesting, as they are simultaneous hermaphrodites, possessing both male and female sex organs. At different times during spawning they can function either as male or female. Always keep this species in a small group.

229 Rabbitfish are less aggressive than tangs

For a peaceful alternative to the territorial tangs, a rabbitfish (*Siganus* spp.) is a good choice. These algae-grazing fish are often found in large schools or in pairs in their natural environment, and they are far less aggressive than their close relatives, the tangs or surgeonfish. Some species have venomous spines and must be handled with care, but species such as the foxface *(Lo vulpinus)* and magnificent foxface *(L. magnifica)* are excellent species for most aquariums.

230 Playing dead—but the foxface soon recovers

The foxface *(Lo vulpinus)* is easily upset and responds to stress, including being introduced to a new tank, by darkening its body color and/or by adopting an immobile position propped against a suitable rock or crevice. However, this motionless state passes quite quickly and the fish makes a full recovery. Apart from this worrying display, the foxface is a hardy fish for the saltwater aquarium.

231 Pufferfish inflate in response to threat

As a defence mechanism, pufferfish and porcupinefish (*Diodon* spp.) have the ability to inflate themselves with water (not air), thus greatly increasing their size to deter predators. In captivity, inflation is only seen in response to an imminent threat or severe stress. It should not be precipitated by the aquarist merely for its curiosity value. Should you have to catch a pufferfish, use a container big enough to allow space should the fish inflate. It is important that you do not lift the fish out of the water as it may then inflate with air and experience great difficulty in deflating. In this condition it will float on the surface and suffocate if action is not taken. Gently push the fish under the water surface to help it deflate.

232 Good candidates for the fish-only aquarium

Hamlets or sea basses (family Serranidae), which are endemic to the Caribbean, make excellent additions to larger fish-only aquariums. The best species are the butter hamlet *(Hypoplectrus unicolor)*, shy hamlet *(H. accensus)*, blue hamlet *(H. gemma)* and indigo hamlet *(H. indigo)*. All are predatory and will consume small fish and crustaceans.

Tip 230 *Its distinctive facial markings give the foxface* (Lo vulpinus) *its common name.*

Better Saltwater Fish

233 Blue-spotted jawfish are beautiful but expensive

The blue-spotted jawfish *(Opistognathus rosenblatti)* is a stunning species that commands a high price where sold. However, it is one of the more hardy jawfish species and will usually feed readily and settle well into the aquarium. Provide a sandy substrate over 3 inches (7.5 cm) deep to enable this species to construct a burrow. (You can provide a small area of this depth by using a food-grade container such as a sandwich box.) This species shares the jawfish trait of jumping when disturbed, so make sure the aquarium is covered.

234 Sweetlips do not retain their juvenile coloration

Before you invest in any species of sweetlips (family Haemulidae) frequently offered for sale, research their maximum size and, if possible, try to see a picture of them in their adult coloration. Juvenile sweetlips are brightly colored, with an endearing swimming style, yet change significantly over time into relatively drab adults. It is thought that the juvenile's coloration may mimic that of toxic flatworms.

235 Wreckfish are hardy but need plenty of food

Wreckfish (*Pseudanthias* spp.) are the archetypal saltwater fish and frequently available in the hobby. Some species should be avoided, including the stunning purple anthias *(Pseudanthias tuka)*, but most are hardy in a reef aquarium, provided they have sufficient food. In common with many planktivores, they feed constantly in their natural environment, and you must be prepared to feed particulate frozen foods little and often if the fish are to thrive. In the wild, huge schools aggregate over reefs; keep small schools of three or more in the aquarium.

236 Hawkfish are compact but predatory fish

Hawkfish (family Cirrhitidae) are an endearing group of saltwater fish that prove hardy and generally peaceful in the saltwater aquarium. However, they are predatory and will consume any small fish or invertebrate that fits into their capacious mouths. Most species available in the hobby do not grow very large, although one exception is Forster's hawkfish *(Paracirrhites forsteri)* that reaches a substantial 10 inches (25 cm).

Tip 235 *Wreckfish* (Pseudanthias squamipinnis)

Tip 236 *Flame hawkfish* (Neocirrhites armatus)

237 Similar to a hawkfish but not predatory

If you like hawkfish (Family Cirrhitidae) but cannot introduce them into your aquarium because of their predatory habits, you might consider the red-blotched perchlet *(Plectranthias inermis)*. This species is becoming more widely available to hobbyists, and although it commands a relatively high price for such a small fish, it is hardy and peaceful and feeds readily on most aquarium foods. Despite its obvious similarities to the hawkfish family, this species is actually one of the fairy basslets (family Serranidae, subfamily Anthiinae).

238 Unrelated, but they have evolved in the same way

Many fish species demonstrate a phenomenon known as convergent evolution. This occurs when two unrelated species occupy very similar ecological niches. Gobies of the genus *Gobiodon* and the spotted coral croucher *(Caracanthus maculatus)*, a close relative of the scorpionfish, are a good example. Both live in the narrow galleries between the branching arms of stony corals and both have become laterally compressed, with rounded heads and strong pelvic fins used to maneuver through confined spaces.

239 New species to the hobby may not yet have a name

More and more species of fish are being sourced and collected for the saltwater aquarium hobby every year. Some species are so "new" that they have not been formally described by science. However, only a very few are not closely related to other species and by applying existing information about the "old" species to the new ones, you should not be too far off.

240 A new introduction that is a fine aquarium fish

The tiny dartfish (*Tryssogobius* spp.) is a new species, both to science and to the aquarium hobby. Despite its small size (about 1 in./2.5 cm) and delicate appearance, it is a hardy and peaceful species that adapts well to aquarium life. Feed it more than once a day on frozen foods and mix it with other species of small fish. This is a perfect fish for the microreef or nanoreef aquarium, although it will thrive in most situations. Microreefs and nanoreefs are very small aquariums, often established by experienced aquarists or those on a very tight budget. Such systems are by definition only a fraction of the size of a standard saltwater aquarium and can be run from a larger system or established to stand alone. Because small volume means less stability, they often require more maintenance than a larger aquarium.

Tip 240 *Dartfish* (Tryssogobius *spp.)*

241 Stock up with invertebrate herbivores gradually

A good basic strategy is to introduce one invertebrate herbivore per 1 gallon (4 L) of reef aquarium capacity. You do not have to introduce these all at once. Build up your population gradually. In fact, later on in your reef's life, if you go out to buy corals but can't find anything you like, use the money to acquire more herbivores. They're not a one-time purchase; you'll need to top up occasionally to make up for losses due to natural causes, accidents and possibly predation.

242 "Turbo snail" is a catchall term for various species

"Turbo snail" is used to describe any of four species of snail, all with a conical shell. *Astraea* species often have raised ribs running down the shell, producing a starlike appearance when viewed from above. *Tectus* and *Trochus* tend to have a fairly regular pyramid shape, while in *Turbo* species the shell has more pronounced whorls. If you can, buy a mix of the different species—the greater diversity the better.

Tip 242 *Turbo snail* (Turbo brunneus)

Tip 243 *Red-legged hermit crab* (Paguristes cadenati)

243 Red and blue hermit crabs reach the same size

The two most popular hermits are red-legged and blue-legged. The red-legs tend to be bigger and cost more than the blue-legs. The most cost-effective approach is to buy blue-legs at the start on the basis that you want a good number of hermits in the tank and you will get more for your money with the blues. As for size, they will grow to the same size as the reds.

244 Which herbivores should I buy and how many?

The usual herbivores to start off with are turbo snails and hermit crabs. The number you need to start off with will depend on the size of your tank. The proportion of snails to hermits will depend on the sort of nuisance algae (if any) you have growing in your tank. If there's no particular algae problem, go for equal numbers of snails and hermits. If you have a hair algae problem, opt for more hermits than snails (about three to two). If algal film is the problem, reverse this (about three snails to two hermits).

245 Hermit crabs may molt soon after introduction

Don't be disheartened if you see what looks like the dead body of a hermit crab a day or so after introduction. It is probably a discarded molt. Like all crustaceans, hermit crabs undergo periodic molting. Shedding its exoskeleton makes way for a new one, allows for growth and provides an opportunity to regrow any limbs or other appendages that have been lost. The new exoskeleton, soft at first, expands and then hardens.

246 Changes in water conditions may trigger molting

Often, the process of molting seems to be triggered by a change in water conditions, so it is not unusual for it to happen soon after introduction. During molting, and for a short time afterward, the crab or shrimp is vulnerable to predation. A crustacean that has recently molted may disappear for a time while the new carapace hardens. Sometimes something can go wrong during the molt and the animal dies.

247 Make sure you have some spare shells handy

Hermits can be very entertaining with their habit of trying on new shells, giving them a quick run around the block, then returning to their old shell if "not completely satisfied." Beacause of this, you should have spare shells of progressively larger sizes for them to move into as they grow. In an aquarium with no spare shells, fighting between hermits can ensue over the possession of suitably sized, desirable shells—or snails may be killed for their shells. Good aquatic dealers will supply extra shells on request when you buy the hermit crabs.

248 Other suitable crabs for the aquarium

Alternative algae-eaters to add are Sally Lightfoot crabs (*Grapsus grapsus*) and emerald crabs (*Mithraculus* spp.). Sally Lightfoot crabs may be a more interesting addition; they are always active and on view. Emerald crabs tend to be more retiring and less visible but still worth adding, perhaps at a later time, just to add to the variety of herbivores.

Tip 248 *Sally Lightfoot crabs* (Grapsus grapsus) *will be highly visible in the aquarium.*

249 Cleaner shrimp are a valuable addition

The cleaner shrimp *(Lysmata amboinensis)* is the best choice of shrimp for the smaller aquarium, unless you are particularly keen on one of the other species. Kept as a pair, they will set up a cleaning station offering their services to your fish and cleaning them of parasites, dead skin, etc. They regularly produce eggs and larvae that make great natural plankton for the other reef inhabitants. They are nearly always on display and when you are working in the tank will treat you just like a large fish, inspecting your hands in search of anything edible.

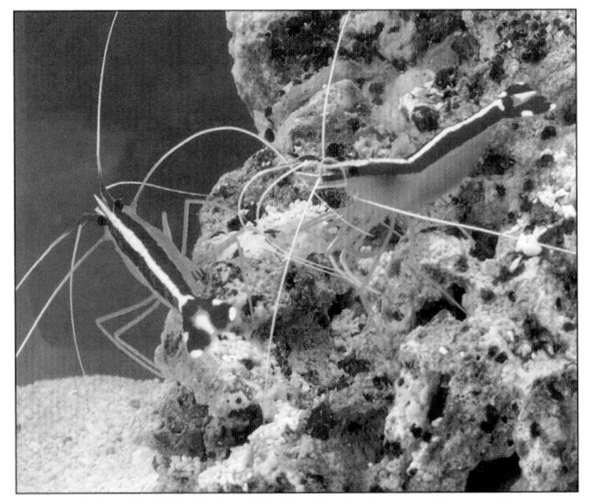

Tip 249 *Cleaner shrimp* (Lysmata amboinensis)

250 Peppermint shrimp can deal with glass anemone

The peppermint cleaner shrimp *(Lysmata wurdemanni)* is similar to the cleaner shrimp but seems to do better in larger social groups. It will mix safely with cleaner shrimp, but may be a bit more secretive. These shrimp are useful if you need to get rid of an infestation of the pest glass anemone (*Aiptasia* spp.). Beware of being incorrectly sold candy, or dancing, shrimp *(Rhynchocinetes durbanensis)* as peppermint shrimp. These are not reef-friendly and eat polyps.

251 Scarlet shrimp are more aggressive than other shrimp

With the scarlet shrimp *(Lysmata debelius)*, also called blood or fire shrimp, there is more potential for aggression; don't mix them with other shrimp species in smaller tanks. They are best kept in small groups. They show similar breeding behavior, but will be more secretive than cleaner and peppermint cleaner shrimp.

252 True pairs of boxing shrimp become devoted

The boxing shrimp *(Stenopus hispidus)* is potentially aggressive, so do not mix it with other shrimp species in smaller tanks. Only keep one shrimp in an aquarium measuring less than 60 inches (150 cm) unless you are certain you have a true pair. If you do have a pair, they'll be secretive except when food is offered. They will breed regularly and demonstrate devoted behavior toward each other.

Tip 250 *Peppermint cleaner shrimp* (Lysmata wurdemanni)

Tip 255 *Yellow polyps—a good choice for the beginner.*

253 Keep anemone shrimp in an invertebrate-only tank

Anemone shrimp (*Periclimenes* spp.) are small shrimp, possibly best kept in an invertebrate-only tank with appropriate species of host corals and anemones, unless you are very sure of the fish species you intend to keep with them.

254 The difference between soft and stony corals

Soft and stony corals belong to the class Anthozoa, but soft corals are in the subclass Octocorallia, so-called because each polyp has eight tentacles. Soft corals in the order Alcynoacea have rubbery tissues with small skeletal structures (spicules) made of calcite, which can be used to identify individual species. Stony corals belong to the subclass Zoantharia, along with the zoanthids Actinaria (anemones) and Corallomorpharia (mushroom corals). Stony corals (order Scleratinia) build a stony skeleton, which forms the fabric of coral reefs.

255 Yellow polyps make a good first coral

If you are a little nervous about introducing your first coral, choose yellow polyps (*Parazoanthus* spp.). They are hardy, colorful and above all cheap! Although they are a photosynthetic coral, they do not depend on really bright light. They appreciate additional feeding with adult-size brine shrimp (*Artemia* spp.) or mysis shrimp (*Mysis* spp.), about every second or third day at first, increasing to daily feedings as the reef matures.

256 Other starter corals for new reefkeepers

Other corals to consider at this stage are the various species of mushroom corals (order Corallimorpharia), zoanthid polyps (order Zoanthidae) and star polyps (*Pachyclavularia* spp.). All these can grow successfully at the lower light levels used during the early life of your reef aquarium. They are all hardy, with little in the way of specialized feeding requirements, although some mushrooms and zoanthids will, like the yellow polyps, appreciate supplemental feeding.

Soft coral choices

257 Many interesting and easy soft corals tend to get lumped together as leather corals, finger corals, etc. Good choices here include the following:

Sarcophyton spp. (toadstool corals) A hardy coral of distinctive shape that can grow quite large and often makes a spectacular impact as a specimen coral.

Sinularia dura (cabbage or flower leather corals) A hardy coral that tends to form leaflike colonies with few polyps. One of the few sinularias, it is possible to identify to species level.

Sinularia spp. (finger corals) Hardy corals with long fingerlike lobes. Many different species of varying forms and colors are available in the saltwater hobby.

Klyxum spp. (colt or pussy corals) Another finger coral that tends to grow quite large. Slimy to the touch.

Cladiella spp. (finger leather corals) Slimy corals of various forms. One species is moundlike, resembling a cauliflower, while another species has the form of a finger coral but is identifiable as a *Cladiella* due to the way it appears white when the polyps retract.

Toadstool coral (Sarcophyton *spp.)*

258 Soft corals and anemones should be attached to rocks

When buying soft corals or anemones, always make sure they are firmly attached to a piece of rock. Do not buy specimens pried from their previous surface; they may have been physically damaged in the process. Sometimes, such unattached specimens refuse to reattach to rocks once you get them home.

259 Frags are a cheaper way of buying corals

Despite often carrying frightening price tags, corals need not be too expensive. Look for "frags"—fragments of coral that have been adhered to rocks (usually using epoxy glue) and grown in a coral nursery until they are of marketable size.

Tip 259 *A typical frag of acropora coral.*

Tip 260 *Coral being double-bagged in the store.*

260 Corals should be bagged up while underwater

Corals are packed in the same way as fish, although they should be at least double-bagged to prevent the weight of the rock and water puncturing the bags. In the store, make sure your coral is bagged underwater to help preserve any sponges that may be attached to the rock.

261 Prevent corals sinking during acclimatization

When acclimatizing a new coral in the reef aquarium, you will find that it is difficult to float a rock! Use plastic clips to hold the bag in place just beneath the water surface. Try clothespins (with no trace of soap!) or the plastic G-clamps modelers use.

262 Adding new corals to the aquarium

When adding new corals to your display, always start them off low down in the tank to help avoid "photo shock." Gradually move them up into their final position in a couple of steps spaced three or four days apart.

Tip 262 *Slowly acclimatize new corals to strong light.*

263 Mushroom corals will thrive in lower light levels

Mushroom corals tend to come from deeper water, so can do well in aquariums with lower levels of lighting (by reef standards), or where the lighting has a greater bias toward the bluer wavelengths. The increased amount of blue light helps to replicate a deeper water biotope and brings out the corals' fluorescent pigments.

264 Leave plenty of space between corals

Always leave a generous gap between new corals and any already established in the aquarium. Bear in mind that most corals of different species are in a constant state of war and can damage each other if placed close enough to touch. Remember—your corals are all going to grow!

265 Zoanthid and star polyps soon colonize rocks

By choosing corals carefully, you can create the impression of a well-populated reef in the smaller aquarium without spending excessive sums of money. In good aquarium conditions, the various species of zoanthid polyps and star polyps all reproduce themselves reasonably quickly, increasing the size of the colony by growing onto and over adjacent rocks.

266 Positioning zoanthids for maximum impact

With zoanthids, you can often buy a specimen in which the colony is growing over a small group of rocks. You can quite easily tease them apart to produce a number of smaller colonies. Space out these rocks around the reef to give the impression of more life, or position them in one location to increase the apparent size of the colony.

267 Increasing star polyp colonies artificially

Star polyp (*Pachyclavularia* spp.) colonies may also be increased right from the start. The coral grows as a mat and often pieces of this mat can be teased apart, loosely attached to small pieces of rock with rubber bands and then used in the same way as described for zoanthids.

Tip 265 *Star polyp (*Pachyclavularia *spp.) colonies soon spread in the aquarium.*

Tip 269 *A well-established reef aquarium with thriving stony corals.*

268 Corals are load-free additions to the aquarium

As soft corals put no significant load on the system—in fact each one comes with its own little bit of extra filtration in the form of the piece of rock to which it is attached—you can add extra corals at will. Most people do not have bottomless pockets, so in reality this means adding one or two corals every one or two weeks or perhaps once a month. There are several varieties to choose from that are not too difficult to keep. (See also Tip 257.)

269 Don't add stony corals to the tank right away

It is a good idea to wait six months or so before starting to introduce stony corals. Stony corals have higher requirements for calcium and carbonates, so you will be far better prepared to look after them when you have had a bit of practice learning to measure calcium and carbonate levels, and supplementing accordingly, with less demanding species of soft corals.

270 Undefended substrate is a haven for pest algae

In the early days of a new reef, you will have a lot of what is known as "undefended substrate"—naked areas on the rock where nothing is growing. These areas are where pest species of algae are likely to take hold. If small pieces of coral, calcareous algae (the pink stuff that is hard to the touch) or other desirable macro-algae are growing on your rock, these should be safe from colonization. If you have built up the reef so far with well-covered live rock, try to fill in any bare patches. You can do this by adding more live rock of aesthetically pleasing shape and by continuing to add corals.

271 What are the zooxanthellae I keep hearing about?

Zooxanthellae (*Zooxanthella* spp.) are single-celled algae that live within the tissues of corals. Like other plants, they use the energy of sunlight and carbon dioxide to build basic sugars in the process of photosynthesis. The energy foods they produce are used by the host coral, which provides the ideal habitat for the zooxanthellae to flourish.

Tip 272 *Single species of coral growing rampantly in the Maldives.*

272 Giving corals room to grow prevents conflict

Bear in mind that corals will grow over time, so unless they are of the same species, keep them spaced well apart. Spacing out the corals limits direct contact and the risk of damage. Large clumps of a single species of coral can look very natural and make for a striking display.

273 Corals are beautiful but deadly animals

Corals have evolved to fight for space on the reef and have a number of strategies to enable them to win out over their competitors. These include: direct action by contact using stinging tentacles; contact at a distance by firing out long-distance nematocysts (stinging cells), contact at a distance through discarding toxic or nematocyst-laden mucus, and chemical warfare through the production of toxic compounds (allelopathy). These compounds can be soluble in water and have the potential to affect any other coral in a closed system.

274 Stinging tentacles can inflict damage from a distance

Some stony corals, particularly *Euphyllia* species, can be some of the most strongly stinging corals you will encounter. They are capable of developing long sweeper tentacles that can sting rival corals from a distance. You will also need to make provision for the calcium requirements of these corals, and they will appreciate occasional feeding—as a guide try using mysis-sized particles of food.

275 Why do some of my corals look brown?

Brown coloration in corals can indicate a high density of symbiotic zooxanthellae. This can occur when there is insufficient light for the coral, and the number of zooxanthellae proliferate to provide more food for the coral. As light levels increase, the brown coloring fades and is replaced by brighter, more attractive colors generated by ultraviolet light protection pigments.

276 Some of my corals don't seem to be thriving. Why?

The ad hoc assortment of corals that often make up reef tank displays could go some way toward explaining why some species of coral never seem to thrive in a particular reef setup. The constant warfare between corals can lead to stress and disease, so unless proper measures are taken it seems inevitable that some corals will suffer.

277 Not all stony corals need strong light

Many stony corals of the genera *Euphyllia* (such as anchor corals), *Physogyra* and *Plerogyra* (bubble corals) with a darker coloration are more likely to have been collected from deeper water or more turbid lagoonal conditions and are well suited to lower light levels. Therefore, they do not require metal-halide lighting in the aquarium. Those with bright fluorescent colors will probably have come from more brightly lit areas and will appreciate higher levels of lighting.

278 There are corals for all lighting levels

Faster-growing corals can overgrow rivals, depriving them of light and thus starving them to death. Taller-growing corals can also end up shading out lower-growing ones. By careful choice of species, however, especially in tanks with higher lighting levels, this can work to your advantage, enabling corals with lower light requirements to thrive in the same display.

279 Avoid species that are difficult to feed

Generally speaking, nonphotosynthetic gorgons and nonphotosynthetic corals, such as the brightly colored *Dendronephthya* species, should be avoided. The amount of food required to keep some of these corals alive can present you with some serious challenges, both in providing the food and in dealing with the consequential potential for pollution. These corals are best left to advanced aquarists prepared to learn how to maintain them.

Tip 279 *Gorgons* (Dendronephthya *spp.) present a challenge, even to advanced aquarists.*

Better Invertebrates

280 Well-fed sun corals make a fine display

An exception to Tip 279 can be the sun corals (*Tubastraea* spp.). If you are prepared to feed these corals one small piece of food per polyp on a regular, preferably daily basis, you will be rewarded with a magnificent display of color, and possibly even find the corals reproducing in your reef aquarium.

Tip 280 *The sun coral* (Tubastraea *spp.) is well named.*

281 Make sure your rockscape is anemone-friendly

In an aquarium housing anemones, avoid building a rock wall along the back glass of the tank. If your anemone goes wandering, there is the danger of it ending up behind this rockwork and starving to death due to lack of light. A better approach is to build up one or two rock "bommies" (coral outcrops), or replicate a small section of a patch reef in the center of the aquarium.

282 Keeping anemones presents a challenge in the aquarium

Anemones, particularly the host anemones that clownfish associate with, are not the easiest of animals to look after successfully. They have high light requirements, do not necessarily mix well with corals, eat fish and can wander around. Most anemones probably survive for less than six months in captivity.

283 And yet ... some anemones are easier than others

To increase your chances of success with anemones, only attempt to maintain the bubble-tipped anemones *(Entacmaea quadricolor)* or the long-tentacled sand anemone *(Macrodactyla doreensis)*. Restrict yourself to keeping only one species of anemone per tank, and design the aquarium with anemones in mind. If you take good care of your bubble-tipped anemone, keeping it safe from predation, well fed and in good water conditions, it may well reward you by dividing, presenting you with an extra anemone to increase your display.

Tip 283 *Bubble-tipped anemone* (Entacmaea quadricolor)

Tip 282 *Even the anemones for clownfish may not be easy to maintain.*

284 Anemones are vulnerable to heaters and pumps

Two big dangers to anemones are in-tank heaters and pump intakes. If you must have a heater in the aquarium, be sure to use a heater guard. If you are keeping an anemone, put foam prefilters on any pump intakes to prevent the anemone being pulled into the pump and pureed.

285 Anemones do better on medium-sized foods

Do not feed overlarge particles of food to an anemone; this can result in partially digested food being regurgitated into your reef, polluting the aquarium. A good size for food items would be the same as adult brine shrimp or mysid shrimp. In the wild, fish would intercept larger food particles before any anemones could consume them.

286 Anemones have a lethal armory

When adding anemones to the aquarium, bear in mind that not all like to be in close proximity to others. Don't mix different species of anemones together, as they are quite capable of waging war at a distance on alien species by releasing nematocysts (stinging cells) into the water in the same way as corals.

287 Intertidal snails can become marooned

Watch out if you experiment with adding some of the newly available *Nerita* species snails to your reef. Although they are mostly useful herbivores, some species are intertidal and need to be able to leave the water. This can lead to problems with the snails drying out and dying when the tide does not come back in; they may even end up on the floor.

Better Invertebrates

288 Heading for a fall—keep clams secure

Clams and other bivalve molluscs are capable of movement, so make sure they are set securely in place. If a clam high up on the rockwork becomes dislodged and falls mantle-down onto a coral, it can prove fatal for the clam. Clams will benefit from additional nutrition in the form of phytoplankton and fine zooplankton, but do not feed them larger particles of food.

289 Featherduster worms will thrive in good conditions

A potential species for lower-lit areas of your reef are the featherduster worms (*Sabellid* spp.). These are not quite as easy to keep as some authorities indicate, but they can do well in a mature tank with additional feeding of phytoplankton and nanoplankton. Partially bury the tube in the substrate to help protect these elegant creatures against predation.

290 Echinoderms need a mature aquarium

New aquariums do not contain enough food to sustain certain animals, particularly echinoderms such as sea stars. Even when mature, smaller reef systems will only produce limited amounts of food. So beware: do not add more echinoderms than your reef can support.

291 Sea stars and brittle stars require special care

Try to avoid exposing sea stars and brittle stars to the atmosphere, because this can lead to disease in some species. Take plenty of time when acclimatizing these creatures, as they are very sensitive to changes in salinity (as are all echinoderms). At least once a week, supplement the diet of any sea star with food suitable to that species, otherwise death by starvation is inevitable.

Tip 289 *Featherduster worms are ideal for dimmer areas of the tank.*

Tip 292 Diadema *sea urchins have sharp spines!*

292 Sea urchins can sting the unwary aquarist

Invertebrates can present certain dangers to the aquarist. The most common painful experience will probably be the result of impaling yourself on the spines of a sea urchin. Always know where potentially dangerous animals are lurking before putting your hands into the tank.

293 Netting and transporting sea urchins safely

Urchins may present problems when being netted or transported. Use a 32-ounce or 64-ounce (1 or 2 L) food-safe kitchen jug to capture the animal and then transfer it—underwater—into a food-safe container for transportation.

294 Some invertebrates need more acclimatization time

Certain invertebrates, notably echinoderms and molluscs, can be very sensitive to changes in salinity and should be acclimatized over a period of time—up to two hours in the case of sea stars.

295 Never allow sponges to dry out—air can kill them

Keep sponges immersed in water at all times to prevent air being trapped inside the animal. Trapped air is a major cause of death. Avoid placing sponges in areas where they can become smothered by detritus. Blue and purple sponges have symbiotic species of cyanobacteria that require high levels of lighting.

Tip 295 *Purple sponges (*Haliclona *spp.) need bright aquarium lighting.*

296 A healthy reef is home to a wide diversity of creatures

Large numbers of invertebrate species can arrive by accident in association with live rock. By finding out as much as you can about them, you will be able to determine which are harmful—and therefore remove them—and which are beneficial and can be left alone. Generally speaking, the greater the diversity of life in a reef or live-rock-based aquarium, the healthier it will be.

297 There are certain crabs that you should remove

Many species of crab are incidental arrivals on live rock and coral base rock. There are too many species to describe here, but in general, you should remove individuals with red eyes as a matter of course. Another potentially problematic species is the hairy crab (*Pilumnus* spp.), which is covered with beige hairlike projections.

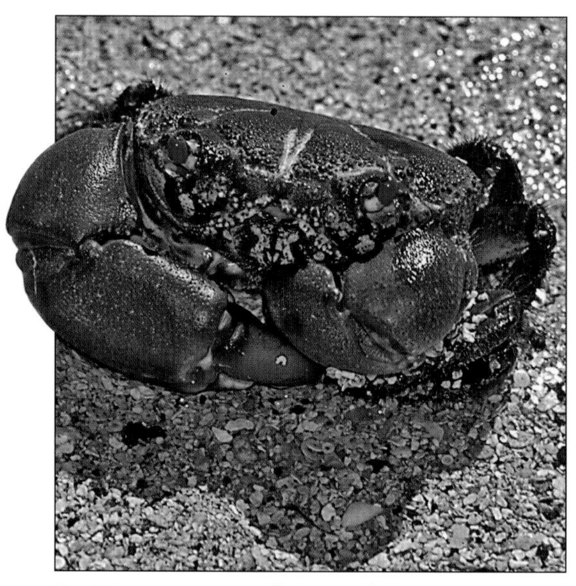

Tip 297 *The red-eyed crab* (Eriphia *spp.) is common on live rock.*

298 Capturing and identifying suspect crabs the easy way

If you are unsure about the identity of a crab introduced on live rock, then try to catch it with one of the commercially available traps or simply use a tall glass baited with a piece of shellfish at the bottom. Place this between some rocks. Once the crab finishes its free meal, it will find it impossible to gain purchase on the smooth-sided glass and will remain at the bottom of the tumbler. You can then have a good look at your captive and decide whether to remove it or not.

Tip 299 *Mithrax crabs are generally beneficial.*

299 Mithrax crabs mostly browse on macro-algae

Most aquariums containing live corals or rock will be home to mithrax crabs (*Mithrax* spp.). These small crabs usually only browse on macro-algae and are therefore generally beneficial. They can be identified by their robust-looking pincers with spoon-shaped ends. They share this trait with the emerald crab (*Mithraculus* spp.) kept by many aquarists to aid in algae control. However, there is hardly a crustacean alive that won't turn down easy protein, so given the opportunity they will also scavenge uneaten fish food and any unwary animals.

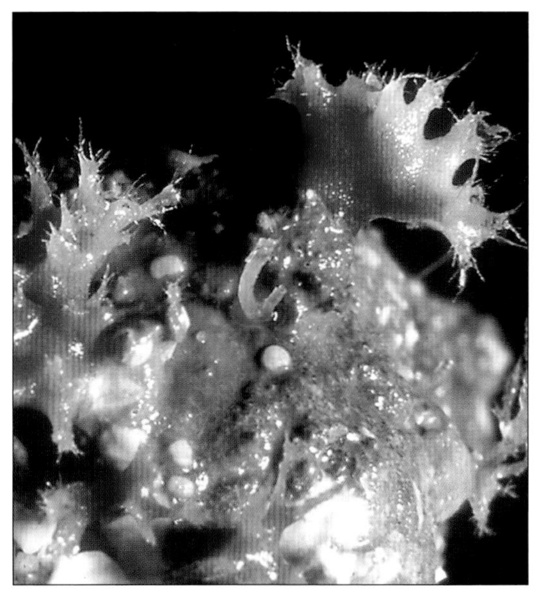

Tip 300 *Forams* (Homotrema rubrum) *on live rock.*

300 Forams and tanidaceans will appear on the glass

When you introduce live rock into an aquarium, you will notice many different organisms appearing on the glass. There will be small single-celled organisms called forams (*Foraminifera*), small antlike crustaceans called tanidaceans and probably many more besides. All will undergo population blooms as the aquarium develops and should be left alone.

301 Amphipod crustaceans are food for mandarinfish

Amphipod crustaceans are small, half-moon-shaped animals resembling sandhoppers. They are very common, particularly in places where detritus accumulates, such as on pump prefilters. They are extremely useful animals in any aquarium and are also the natural food of fish such as the mandarinfish (*Synchiropus* spp.), also called dragonets, scooter dragonets or scooter blennies.

302 Pistol shrimp and mantis shrimp are easy to confuse

Two crustaceans that are often confused are the pistol shrimp (*Alpheus* spp.) and the mantis shrimp (order Stomatopoda). Pistol shrimp use their enlarged pincers to make a sound like breaking glass, but are not usually too problematic in the reef aquarium. However, mantis shrimp are highly predatory and large specimens can actually break glass, hence the confusion. The mantis shrimp should definitely be removed but it makes a wonderful pet when kept in an aquarium of its own.

303 Stomatella are beneficial algae-grazers on the reef

Small sluglike snails are common accidental imports associated with live rock. They are called stomatella (*Stomatella* spp.) and have a habit of shedding their tails as an antipredator response. Although they resemble nudibranch molluscs that can often be harmful in a reef aquarium, stomatella are beneficial algal grazers that aquarists should leave alone.

Tip 303 *Stomatella can safely be left in the aquarium.*

304 The checkered box snail hides in zoanthids

Some predators do not need to master camouflage or ambush techniques to kill their prey. The checkered box snail (*Heliacus* spp.) can often be found nestling snugly in a clump of its favorite food, button polyps (zoanthids). Before you buy them, always check colonies of button polyps for this unsubtle snail. If you discover the snail, remove it. The zoanthids should recover with no adverse effects.

Tip 304 *Remove the checkered box snail* (Heliacus *spp.) from zoanthids.*

305 Dove snails feed on algae on the aquarium glass

The dove snail (*Euplica* spp.) is a useful species, commonly encountered yet free when found on live rock. This hardy snail species seldom reaches more than about a half inch (1 cm) in length and feeds on algae, particularly the type that grows on aquarium glass. It reproduces readily in the aquarium and the eggs hatch into minute replicas of the adults.

306 Many bristle worms perform a useful cleaning function

There are over 9,000 species of polychaete worm currently known. Aquarists often refer to these free-living species as "bristle worms." Only a small percentage are harmful to fish or invertebrates and many actually perform a useful service scavenging on uneaten fish food and detritus. Never handle bristle worms without wearing gloves.

307 Nudibranch sea slugs are masters of disguise

Reef aquarists are likely to encounter several species of nudibranch sea slug, most of which closely resemble their prey. Perhaps the most common is *Tritoniopsis elegans*, which feeds on soft corals from the genera *Cladiella, Sarcophyton* and *Lobophyton*. These coral genera contain some of the most popular aquarium species. If you note any damage to one of these soft corals, check the base for this beautiful nudibranch and remove it immediately.

Tip 307 Tritoniopsis *nudibranch on soft coral.*

308 Parasitic snails feed on blue starfish

The blue starfish *(Linckia laevigata)* is a highly attractive subject for reef aquariums, but when you buy one you might get more than you bargained for. The beautiful yet parasitic snail *Thyca crystallina* is often found on the upper surface of this and other starfish, where it consumes the flesh of its host. Removing it will leave a mark, but the starfish should recover.

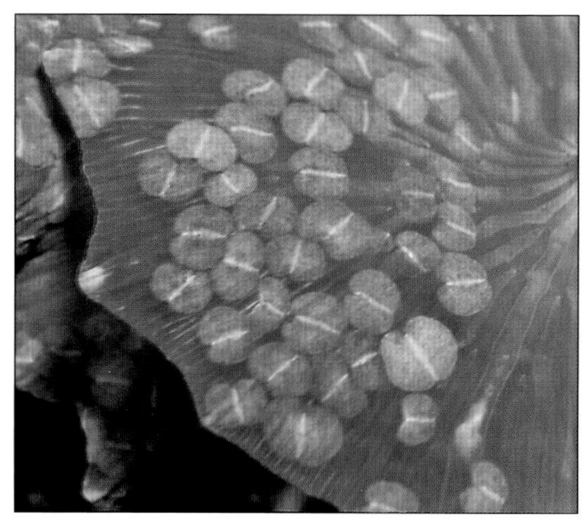

Tip 309 *Planarian flatworms clustered on* Corallimorpharia.

309 Planarian flatworms often die out naturally

The rust-colored planarian flatworm *(Convolutriloba retrogemma)* is a real problem in many reef aquariums, as it is capable of asexual reproduction and houses photosynthetic algae that nourish it. Removing these persistent little pests can be difficult. The nudibranch *Chelidonura varia* consumes them and is often available in the hobby, but will never dispose of every single flatworm. Often the best strategy is simply to do nothing; the population of planarians will dwindle naturally.

310 *Asterina* sea stars do not deserve their bad reputation

Small *Asterina* sea stars, less than half an inch (1 cm) across, are often found in reef aquariums, arriving as hitchhikers on live rock and corals. They reproduce by fission, tearing themselves in two and growing new legs along the split, hence their uneven appearance. There are a number of similar species but identification is difficult. They are scavengers and feed on microorganisms. As they graze on necrotic tissue, it has been mistakenly assumed that they predate on corals.

311 Tiny *Echinometra* sea urchins grow rapidly

Sea urchins of the genus *Echinometra* frequently go unnoticed in reef aquariums due to their small size on arrival (some are only as big as a pea). However, they grow rapidly and have an incredible appetite for calcareous algae. Such large echinoderms are clumsy and will knock over unsecured corals. If you can live with this, then the urchins present few other problems. If not, remove them as soon as you spot them.

312 Peppermint shrimp will remove glass anemones

Glass anemones (*Aiptasia* spp.) are common pests found on live rock and coral base rock. Do not under any circumstances attempt to remove these anemones physically, as any damage can result in the anemone reproducing out of hand. Instead, use one of the commercial preparations that are available, or stock some peppermint cleaner shrimp *(Lysmata wurdemanni)*, which will consume these nuisances.

Tip 312 *Remove glass anemones* (Aiptasia *spp.) with care.*

Better Compatibility

Tip 315 *Think long-term when choosing fish and invertebrates.*

313 Compatibility is crucial for a marine setup to succeed

Compatibility is of the utmost importance when setting up a saltwater aquarium. The aim is to create a harmonious environment, with none of the inhabitants preying on each other. Compatibility problems are not limited to fish. When invertebrates are added to the mix in a reef aquarium there is a whole new set of compatibility issues to consider. Many fish will quite happily devour small invertebrates as part of their omnivorous diet. Occasionally the reverse is true; some invertebrates, including mantis shrimp, swimming crabs and tube anemones (*Cerianthus* spp.), will prey on fish. Giant mushroom polyps or large hermit crabs have been known to take unsuspecting fish.

314 Ask your dealer for advice on compatibility

If you are in any doubt about compatibility issues with a particular fish, consult your dealer before you buy. Take a list of the fish and other species contained in your aquarium and ask whether the fish you are considering is a suitable addition. Even if you do not elicit a straight yes or no answer, you will still acquire some useful information regarding possible conflicts between particular species.

315 Think ahead when stocking your aquarium

Many people stock their aquarium with a view to acquiring a particular species once their aquarium is capable of sustaining it. Try to plan any stocking regime around this species, as careless introductions in the early days of your aquarium can mean that your favorite species must be excluded later on.

316 Some fish cope better than others with passing traffic

Do not think about compatibility simply in terms of the combination of fish that you are likely to introduce and whether they are reef-safe or not. An aquarium exposed to a great deal of disturbance because it is located near a doorway or in a hall might not be suitable for timid fish that are prone to hiding when disturbed. However, many more tolerant species will rapidly adjust to passing traffic and become that bit hardier as a result.

Tip 313 *Mantis shrimp will prey on fish.*

Tip 319 *The powder blue tang* (Acanthurus leucosternon) *can prove problematic in the aquarium.*

317 Close relations may view each other with suspicion

On the whole, saltwater fish are a cantankerous bunch, especially in the confines of captivity. Tension and aggression will undoubtedly be heightened between fish of a similar size and body shape, or between fish with a similar pattern of markings. Also bear in mind that the more closely two fish are related, the greater the likelihood that they will not enjoy each other's company.

318 Fish won't always behave as you expect them to

Individual fish can defy even the best advice based on years of experience. Most dealers will be able to give you examples of rogue elements within a species more usually known for their benign and peaceful characteristics. Retailers normally accept such fish back from the hobbyist, but blame cannot be fairly apportioned in these circumstances. The attitude of experienced hobbyists when faced with seemingly abnormal behavior from their fish is understandably, "C'est la vie!"

319 Some fish don't merit their "good" reputation

Some fish manage to avoid the "aggressive" label and acquire a laid-back reputation that is far from justified. One example is the powder blue tang *(Acanthurus leucosternon)*. Very few books hint at the potential damage this fish can and will inflict on more peaceful tankmates. Seeking out the advice of fellow hobbyists can go a long way to avoiding problems.

320 Compatibility can mean keeping species together

When we think about compatibility we tend to think of avoiding one species that could be in conflict with another. However, many species are completely compatible to the point of being inseparable. Some fish team up with another individual because the foraging behavior of one species can locate food for the other. Wrasse will often follow triggerfish and parrotfish that, in the course of their foraging, might flush out small invertebrates for the smaller fish. Such relationships are common in the natural world.

Better Compatibility

321 Feeding habits play a part in compatibility

When considering whether a fish is reef-compatible or not, it is sometimes possible to distinguish between species that are safe with hard or soft corals. In certain species, this distinction can be narrowed down even further. For example, the masked butterflyfish *(Chaetodon semilarvatus)* is unlikely to bother soft corals of the genera *Cladiella* (finger leather corals) or *Sinularia* (finger corals), yet it includes other species of soft coral in its natural diet.

Tip 323 *Yellow and purple tangs in the same aquarium.*

Introducing territorial species at the same time

When attempting to introduce potentially territorial species into the aquarium, it is often best to add them simultaneously, provided the system can take the strain of adding two or more fish at the same time. By adopting this strategy you can be confident that aggression will be kept to a minimum, as the potential protagonists will meet before territories have been established.

Tip 321 *Masked butterflyfish* (Chaetodon semilarvatus)

Introducing territorial species one after the other

If you plan to introduce two closely related and territorial species but cannot do it simultaneously, then try adding individuals of differing sizes. Put a small individual of one of the species in first, followed by a smaller representative of the second species. An example of a situation where this would work well is with *Zebrasoma* tangs, such as the yellow tang *(Z. flavescens)* and the purple tang *(Z. xanthurum).*

324 Even compatible additions can face aggression

Despite your best efforts to reduce the risk of aggression, even compatible species can take some time to get used to one another. Following the introduction of a new fish, the existing residents will often gang up on it and display their aggression. Observe what is happening from a distance and be ready to intervene should the situation escalate into all-out war. The newly introduced fish is quite vulnerable at this stage, as it is often also adjusting to the change in water parameters. The situation usually calms down after, at most, a few hours, provided the new fish introduction was well researched.
(See also Tip 182.)

Tip 325 *Dottybacks (this is* Pseudochromis diadema*) may threaten ornamental shrimp but are harmless to corals.*

325 Invertebrate and coral compatibility are different

When trying to establish whether a particular fish species is suitable for introduction into a reef aquarium, remember that there is a difference between invertebrate compatibility and coral compatibility. For example, dottybacks (*Pseudochromis* spp.) will not harm corals, yet can present a threat to ornamental shrimp.

326 Give corals plenty of space in the aquarium

Natural reefs are a war zone. Species of coral compete for the available substrate through chemical warfare, overwhelming their enemies through fast growth and by physically stinging them to death. Some species of coral even use their digestive organs to devour the opposition. Allowing specimens space to grow will minimize the impact of any interspecific competition. Think long-term and allow for ample growth when positioning corals.

327 Anemones can pose a threat to fish

If you keep anemones, be aware that they constitute a risk to any fish kept in that aquarium, including—when carpet anemones (*Stichodactyla* spp.) are taken into account—clownfish. If you do your research you will become aware that there are many fish and invertebrates best suited to some form of species tank.

328 Find the right anemone for your clownfish

By pairing a species of clownfish (also called anemonefish) with the species of anemone that it associates with in the wild, you will go a long way to ensuring that the relationship starts well. Some species of clownfish will accept any anemone offered, but others are far more particular. If you have an unfussy clownfish or two, then you are fortunate, as you can choose the anemone best suited to your aquarium. (See also Tip 160.)

329 Clownfish evict other occupants of their anemone

Clownfish will rarely tolerate another animal living in their anemone. Introducing a pair of clownfish into an aquarium containing an anemone hosting an anemone crab (*Neopetrolisthes* spp.) or anemone shrimp (*Periclimenes* spp.) is likely to result in a homeless crustacean that, once separated from its protector, could fall prey to another fish.

Tip 329 *Anemone shrimp* (Periclimenes *spp.*)

330 Choose starfish species with care

Certain species of brittle or serpent starfish should not be introduced to an aquarium that is home to small fish. For example, green brittle starfish of the genus *Ophiarachna* deliberately catch and eat fish that shelter beneath them. However, many other species of brittle and serpent starfish are very useful as scavengers in saltwater aquariums, provided they do not grow too large. Examples include *Ophiolepis* spp. and *Ophiothrix* spp.

331 Are sea urchins suitable for a reef aquarium?

It is very tempting to introduce small sea urchins into your reef aquarium, particularly long-spined individuals of the genus *Diadema*. Notwithstanding the threat their sharp spines pose to the aquarist, these clumsy animals, with their huge appetite for calcareous algae, rapidly attain a spine diameter in excess of 24 inches (60 cm). Before you buy them, carefully consider whether your aquarium will accommodate such creatures in the long term.

332 Boxfish do not welcome attention

The slow, cumbersome nature of boxfish, such as the cowfish *(Lactoria cornuta)* or the ever-popular humpback turretfish, also called hovercraft or helmet cowfish *(Tetrosomus gibbosus)*, leads to an often overlooked problem in captivity: they can be driven to distraction by the unwanted attention of a cleanerfish such as a cleaner wrasse (*Labroides dimidiatus*) or neon goby (*Elacatinus oceanops*). Never house these species together, as you risk incurring a toxin-induced wipeout. Indeed, to avoid such a disastrous outcome, there is a good argument for keeping boxfish and cowfish only in species tanks.

Tip 330 Ophiolepis *spp. starfish are useful scavengers.*

Tip 333 *The lemon goby* (Gobiodon citrinus) *and other* Gobiodon *species enjoy a natural defense against predation.*

333 Certain gobies have a defense against predation

Gobies from the genus *Gobiodon* can be mixed with most other smaller species of saltwater fish as they are covered with distasteful mucus that other fish, including potential predators, find abhorrent.

334 Keep just one species of sand-dwelling goby

There are many species of goby adapted to live on sandy substrates. Try not to mix different species if possible. Although there are notable exceptions, such as some species of *Stonogobiops* shrimp goby, most sand-dwelling species are quite possessive about their beachfront property and will defend it aggressively.

335 A predator aquarium needs a lot of research

Do not be tempted to establish a "predator" aquarium without much research. Despite being venomous, many a slumbering lionfish has fallen victim to a hunting moray eel, and any difference in growth rate between individual predators can mean that even if harmony exists in the aquarium it might be only temporary.

336 Make sure you know how large a predator will grow

You can keep predators quite happily in a community setting. However, if a fish is small enough to be swallowed by a predator, then you can virtually guarantee that it will be. This may not be an issue when the predator is small, but ensure you know its growth potential, otherwise you have a recipe for disaster.

Better Feeding

337 Look at the position of a fish's mouth

Do not give fish food they do not want or cannot eat. Judge by the position of their mouths where they naturally eat. Top mouth = surface feeder; terminal mouth = midwater; underslung mouth = bottom feeder or grazer. Long-snouted fish, such as butterflyfish, are "crevice-pickers"; barbel-equipped goatfish are substrate-sifters.

Tip 337
The lyretail grouper's (Variola louti) *mouth is a clear indication of its predatory nature.*

338 Find out about your fish's feeding strategies

Get to know the different types of feeding mechanisms employed by saltwater fish so you can correctly feed your stock. Your fish may be grazers of coral polyps, algae growth and sponges, or they may pick at the myriad of tiny life-forms found within the crevices of a reef. Fish, both large and small, may be predators on other life-forms, ranging from plankton to other fish. Many fish are omnivorous and will eat whatever the opportunity presents. Some fish, such as triggerfish, take their omnivorous nature to the extreme and even try to sample aquarium equipment, but not all fish are so accommodating. Some are specialized feeders, requiring large quantities of tiny live food, while others may be so highly specialized that they only eat particular sponges.

339 A balanced diet is essential for fish health

Providing a feeding regimen that is acceptable to saltwater fish, while providing them with all their nutritional requirements, is fundamental to their health and survival. Nutrition has a direct impact on the appearance, color, growth rate, disease resistance, reproductivity and general well-being of your fish. Their diet must provide the essential components of proteins, fats, carbohydrates, minerals and vitamins. Do not use fish food aimed at the freshwater or coldwater market, as it may contain too much fat. Saltwater fish need a high-protein (40 to 60 percent) but relatively low-fat (5 to 10 percent) diet to thrive. Excess in a dietary component can be as damaging as a deficiency. Use a variety of foods to meet these demanding nutritional requirements.

Defrosted frozen foods.

Saltwater mix
Natural saltwater invertebrate and fish meat.

Shrimp
Many wild-caught foods are irradiated.

Whole cockle
Accepted by many saltwater fish.

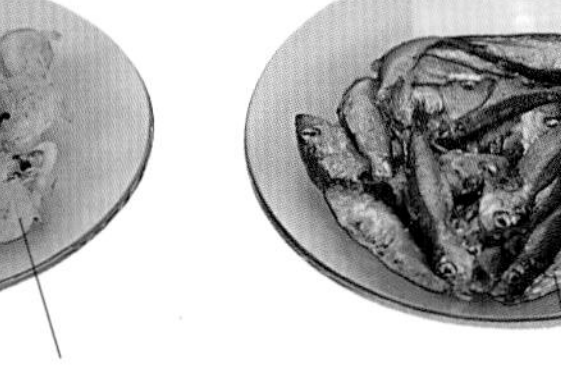

Krill
Nutritious food for larger fish; break up for small fish.

Small fish
A "one-gulp" food for larger fish.

340 Frozen foods are best for saltwater fish

Frozen foods are by far the most popular and appropriate choice for most reef fish. There is a good range available from various sources, including foods formulated for difficult feeders, foods enriched with HUFAs (highly unsaturated fatty acids) and foods that have been irradiated to ensure they are parasite-free. Be aware that frozen food can show signs of bacterial contamination within five minutes of defrosting.

Frozen foods are sold in single slabs. Simply break off pieces as required.

Allow frozen foods to thaw out before feeding them to prevent the fish eating ice.

341 Store frozen food in sealed containers

Keep your frozen food in a sealed container to avoid it drying out (freezer burn). This will also prevent unpleasantness when other family members discover what you are storing in the freezer! Label containers clearly.

342 Some frozen foods produce an oily film

Some frozen foods may taint the aquarium water by producing an oily film on the surface. Remove this by drawing a sheet of absorbent paper across the water surface. Always rinse frozen foods before use; this may help to defrost them and also removes any oiliness.

Tip 343 *Live brine shrimp is a good first food.*

343 Is brine shrimp a suitable food for saltwater fish?

Brine shrimp (*Artemia* spp.) is a useful first food for most saltwater fish because they find it very palatable and accept it readily. However, many species will need far more nourishment than brine shrimp can provide, even if enriched. As frozen foods go, mysis shrimp is a much better staple, as it contains a greater percentage of fat and therefore is useful at keeping weight on fish that might otherwise become emaciated if solely fed on brine shrimp.

344 What is meant by gamma-irradiated foods?

Gamma irradiation is a method by which food potentially containing harmful microorganisms can be made safe for fish to consume. Gamma rays are a form of electromagnetic energy, similar to, but more powerful than, microwaves, that can free electrons from atoms (ionization) to disrupt DNA. This kills the microorganisms and prevents disease passing from the treated food to the fish. The processed food is quite safe to handle and does not contain any residual radiation.

Better Feeding

345 Can I feed freshwater live foods to my saltwater fish?

Although most live foods, such as mosquito larvae, daphnia and bloodworms, should live long enough in saltwater to attract the attention of saltwater fish, it is best to feed them foods of saltwater origin.

346 Providing supplementary vegetable foods

Many saltwater fish require vegetable matter in their diet and, while flake and frozen food can be found in vegetable-enriched formats, some fish will require supplementary vegetable food for good health. Saltwater macro-algae are the most natural way to provide the vegetarian option. You can culture it specifically in your display aquarium or grow and harvest it in a separate tank. The algae can be grown as a "lawn" on rocks in the separate tank. Place individual rocks in the main tank for feeding when they have acquired a lush growth.

Flake and granular foods

Brine shrimp in flake form.

High-quality mixed flake formulations contain many of the vitamins and proteins required by the majority of saltwater fish.

Fast-sinking granular foods help to ensure that bottom-feeding fish receive a supply of food.

Tip 347 *Secure a sheet of dried saltwater seaweed with a clip.*

347 Dried saltwater seaweed for herbivorous fish

Herbivores in the reef should have food available all the time, as they need to be constantly browsing. Supplement the algae naturally available in your reef with dried saltwater seaweed (macro-algae). Many varieties are commercially available, which is just as well, as some tangs can have the habit of settling on one type to the exclusion of all others.

348 Offer saltwater algae in a familiar place

If you are bringing in macro-algae raised in another tank, you can place it in a sucker-mounted clip or between the two halves of a small magnetic algae scraper. Keeping the food in one place serves two purposes: the fish know where to find it, and it won't drift around the aquarium only to be sucked into the filter intake.

349 Give your fish a variety of food

Provide your fish with a varied diet to prevent them becoming too set in their ways. Although modern prepared foods contain all the necessary nutrients and vitamins, even slight differences between various manufacturers' mixes will provide an appetizing change for them. When buying frozen foods, buy a different type each week to build up a good range, and then feed in rotation.

350 Phytoplankton is a vital food in the reef aquarium

Phytoplankton is often referred to as being the basis of all life in the oceans and as such is seen as fundamental to today's reef aquarium. When you feed phyto you are in effect feeding your tank. The primary users of phyto are the minor animals, who will then go on to feed your corals and fish by reproducing and providing edible plankton in the form of eggs, sperm and larvae, or by being eaten themselves. Phytoplankton is available either as live phytoplankton or as a preserved product. Live is best as it is non-polluting.

Tip 350 *Live phytoplankton is supplied in bottles and small bags.*

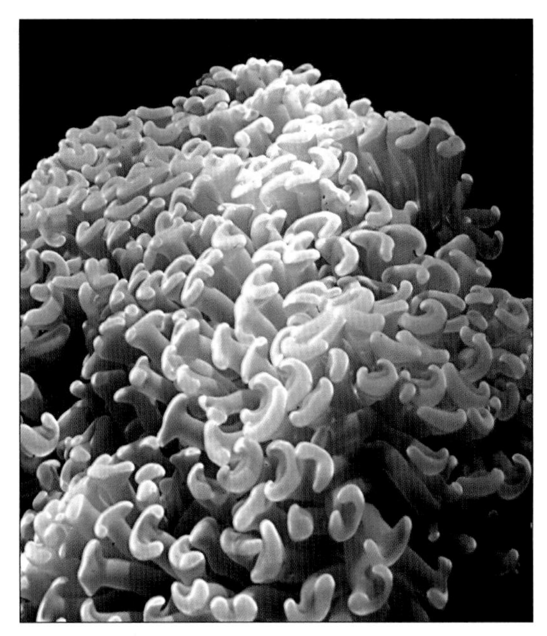

Tip 351 *Corals require sufficient light to survive.*

Remember that light is food for corals

Light is food for algae and also for the zooxanthellae harbored by corals. Make sure you have appropriate levels of light to maintain the animals and plants you want to keep. (See also Tip 271.)

352 Will saltwater fish accept lettuce as a treat food?

All in all it is best to stick to foods of saltwater origin. Over-reliance on terrestrial plant material should be avoided due to the possibility of a build up of oxalic acid over time. (Oxalic acid is the poison that precludes us from eating rhubarb leaves.) Most terrestrial plants, such as spinach and lettuce (blanched in boiling water before use), should be viewed merely as a treat or filler rather than as a main source of nutrition in the saltwater aquarium.

Better Feeding

353 Feed proprietary liquid foods with care

Many of the proprietary liquid foods aimed at corals and filter-feeding invertebrates (bivalve molluscs, sponges, sea squirts, fanworms, etc.) tend to be yeast-based. Be careful when using these foods; although they are undoubtedly nutritious, there is great potential for pollution if used unwisely.

Feeding homemade food with a turkey baster

One good method of feeding corals is to mix frozen foods of appropriate size with phytoplankton, plus a measure of tank water. Stir this until the frozen food has defrosted and then "target-feed" the corals using a turkey baster. Target-feeding means directing food at individual corals, and a turkey baster is an indispensable tool for this job. Do not just blast the food into a polyp, as this may cause it to close up. Instead, gently direct a stream of food toward the coral until you see it capturing its prey. In fact, turkey basters are very useful all-around reef tank tools. In addition to target-feeding, you can also use them to blow detritus off corals or to pick up errant loose polyps.

Tip 356 *A regal angelfish* (Pyloplites diacanthus) *is offered meaty food.*

355 Hard-shelled food helps to keep teeth short

Pufferfish and porcupinefish have formidable teeth that are fused into a beaklike structure. In the wild they use these teeth to make short work of mollusc shells, or even to take bites of hard coral and then consume the softer delicacies contained within. The teeth grow constantly, but are kept worn down to optimum size by the very nature of their diet. In captivity you must provide hard-shelled food to enable these fish to keep their teeth short. Suitable foods such as whole shrimp and cockle in the shell are available in frozen form.

Tip 355 *Pufferfish teeth are fused.*

356 Encouraging predators to accept dead food

Predators sometimes take a while to adapt to an aquarium diet. They appear to be reluctant to strike at and take anything other than live food. The live diet can be accommodated in the form of river shrimp or small freshwater fry, but this will be an expensive, and perhaps unpalatable, option. One method to use in an attempt to overcome this reluctance is to include a meat-based frozen food along with the live food, in the hope that in the feeding frenzy the predator will also take the prepared food. If this fails, the next step is to offer a piece of meaty food in long-handled tweezers or attached to a piece of cotton and patiently trail this through the water in a jerky motion in the hope that the reluctant predator will strike. As you can imagine, this would be quite a performance. It is much easier to make sure your intended purchase is accepting dead food before you acquire it from the dealer.

357 Seahorses require a specialized diet

Seahorses and pipefish require copious amounts of small live food that can only be provided by newly hatched and adult brine shrimp, preferably enriched. Unless you can supply such a diet you will not succeed with these fish and you must avoid them. Buying live brine shrimp will prove expensive if you rely on it as your sole food supply. The alternative is to raise your own from eggs in a homemade system or in a hatchery available from your retailer. Appealing as seahorses may be, they are for the experienced aquarist only.

Tip 357 *Seahorses need plenty of live food.*

358 Some butterflyfish can be very difficult to feed

Many species of butterflyfish are obligate coral polyp feeders, which means that they are known to feed solely on the tissue of these sessile invertebrates. Although a few captive specimens will feed on frozen foods or pick at live rock, the nourishment these provide is not adequate for their long-term survival. Many wholesalers refuse to import such fish, but regrettably they are still collected for the aquarium trade. Species to avoid are the blacktail butterflyfish *(Chaetodon austracius)*, the orange butterfly *(C. melapterus)*, Meyer's butterflyfish *(C. meyeri)* and the eight-banded butterflyfish *(C. octofasciatus)*.

359 Be prepared to provide several meals a day

Certain species of butterflyfish require feeding several times a day. Unfortunately, by the time an aquarist notices that a fish appears to be thin it is often too late. Target-feeding individual fish may be necessary, so think carefully before acquiring butterflyfish. In an aquarium containing boisterous fish that feed readily, more sedate butterfly species are likely to be outcompeted for food. Those with a particular problem include the copperband *(Chelmon rostratus)* and longnose *(Forcipiger flavissimus)*.

360 Making a grazing rock for finicky feeders

Angelfish and butterflyfish spring to mind as the classic fussy feeders. To make a grazing rock for finicky grazers, ingredients such as frozen shrimp, clam, fine particles of mussel, saltwater algae, vegetable flake food and mineral tablets are crushed, mixed and held together with gelatin. The mixture can be frozen and grated for use when required or smeared onto a piece of slate and placed in the aquarium as a grazing rock. Be sure to remove it after feeding.

Better Feeding

361 Give filter feeders a good chance to obtain their food

Unlike fish, sessile invertebrates cannot chase after food and it has to be delivered to them. Turn off filtration systems (but not necessarily water pumps) for a short period while feeding invertebrates with liquid food, otherwise the filter will "eat" the food before the inverts do.

362 How can I persuade saltwater fish to accept a new food?

Changing fish foods can be a tricky endeavor, as many fish take a while to acclimatize themselves to any new, unfamiliar food. Try withholding food for a day or two and then offer small amounts of the new food again.

363 Don't forget the nocturnal feeders

Just as not all fish eat the same thing, not all eat at the same time of day. Some saltwater fish, such as squirrelfish, hide away during the day and become more active at night—when they will obviously be on the lookout for food. Make sure some food is available during the hours of darkness for such species.

Tip 363 *Squirrelfish* (Holocentrus diadema)

Tip 364 *Make sure all the fish get their share at feeding time.*

364 Delicate fish can miss out during a feeding frenzy

Some fish can become very agitated at feeding times. They may snap at anything within reach, or they may zoom around the tank at frenzied speeds as soon as they anticipate that food is about to enter the water. Both types of behavior, while not exactly bullying, can lead to injury or the exclusion of more timid fish at feeding time. Culprits include bruisers such as triggerfish and puffers, or the unsettling, lightning-fast wrasses. Stock the aquarium carefully to ensure that voracious feeders are not kept with delicate species.

365 Buying in bulk can be false economy

It is a false economy to buy larger packs of food just because it is cheaper in bulk. As soon as you open a pack of food it will start deteriorating and certain vitamins will start to break down, resulting in food of poor nutritional quality. Only buy the amount of food your animals will go through in a reasonable time.

366 Keep foods in top condition and follow the instructions

Always read the packaging. Many foods and supplements need to be kept in the fridge to preserve them. Some food products will have best before dates. Pay attention to these and never buy out-of-date food.

367 How much and how often should I feed my fish?

The answer is "little and often." You should net out any food that remains uneaten after two to three minutes. If you need to do this, it means you are either overfeeding or feeding inappropriate foods. Try to feed in the morning, late afternoon and just before lights out. The morning and night feeds are the most important.

Tip 368 *Sprinkle a measured amount of food into the tank.*

368 Take a measured amount of food to the aquarium

Do not take whole tubs of fish food to the aquarium. Sort out how much you need and take just that amount to the tank. Remember that one slight slip of the hand may result in emptying an entire tub of flake food into your bright and sparkling aquarium. The remedial work will take many hours.

Tip 367 *A decorated dartfish* (Nemateleotris decora) *about to capture a fragment of thawed-out shrimp.*

Better Ongoing Care

369 Maintenance is a chance to revitalize the aquarium

Saltwater aquarium maintenance is not as arduous a task as many people imagine. It relies on common sense, plus a sound understanding of saltwater aquarium basics and biological filtration. Look on it as a vital aspect of the hobby, carry out the tasks on a regular basis and regard them less as a chore, more of a pleasure.

370 Partial water changes are a solution to many problems

Biological changes will occur in the aquarium as it matures and ages. However efficient it may be, a filtration system cannot degrade all the waste products generated by the inhabitants. Organic compounds tend to accumulate in the water, as will nitrate in many tanks. A tendency toward a falling pH level is a natural trend, and trace elements will become exhausted over time. The simple process of regular partial water changes is the solution to these problems. Frequent water changes are generally preferred to larger monthly ones. Replacing 5 percent of the water in a 48– to 60–gallon (180–270 L) tank only involves handling 2 to 4 gallons (9–13 L) each week and is an easy regime to follow. Compare this with a monthly change of 20 percent that would require you to reconstitute and handle 12 to 18 gallons (45–68 L) of saltwater.

371 Carrying out a water change is a simple operation

There are many ways of doing a water change. If there are no apparent algae problems, one method is simply to siphon out the water into a bucket using a length of 1/2-inch (12 mm) flexible hose. If there is a biofilm on the surface of the water, scoop it off using a suitable container while changing the water.

372 Air line allows you to siphon out water slowly

If you have an algae problem it can be useful to do the siphoning with a length of air line. Using air line slows down the rate at which you take water from the tank, allowing you to take your time siphoning out pest algae. This is particularly useful if you are trying to eradicate green or red slime algae (cyanobacteria) where the pest species has a tendency to coat the rocks, sand and glass, or if you are trying to rid your tank of the pest photosynthetic flatworm *Convolutriloba retrogemma*. (See also Tip 309.)

Tip 369 *Maintenance is a vital part of the hobby.*

Direct the water from the tank into a bucket on the ground.

Tip 371 *Making a partial water change.*

373 Water changes control the level of trace elements

Water changes also represent a way of introducing your animals' trace element requirements or, alternatively, help to prevent trace elements potentially building to inappropriately high levels.

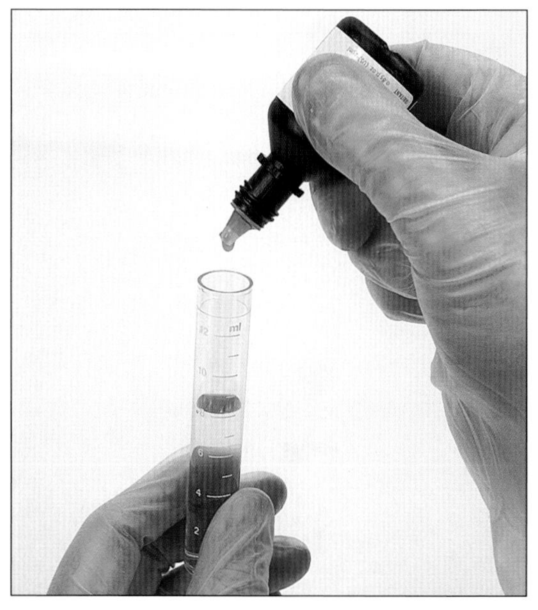

Tip 376 *Testing the water is part of the maintenance routine.*

374 Check the salinity of new water added to the tank

When introducing newly made up water, always double check the salinity. Be careful to pour it into a strong flow and not directly onto one of your corals. This is one of the advantages of a sump, where you can more or less pour in the freshly made up water without having to worry. After you have finished adding the new water, it is worth waiting again about 15 minutes to allow the water to mix. Then check the salinity again in case any adjustments need to be made.

375 Adjusting the salinity of the water in the aquarium

If the salinity is too high, slowly add R.O. water—a maximum of about a quarter of a gallon at a time—being careful not to pour it directly onto a coral. If salinity is too low, mix up a small amount of water at a higher salinity and add that. If there is only a small difference, say specific gravity of 1.024 when you want 1.025, let evaporation do the job for you, topping up with water of the required salinity at a later time to restore the correct water level.

376 Do all the maintenance on your tank at the same time

At the same time as performing the water change, carry out the other maintenance tasks associated with the efficient running of the aquarium. Change airstones, harvest excess algae, clean protein skimmers, check the filters on air pump intakes, check internal power filters and clean the cover glasses. For newly established tanks, check out the main water parameters of salinity and temperature, plus pH, nitrate, nitrite and alkalinity levels. Completing this maintenance at the same time as your water changes minimizes the amount of disturbance to your livestock by reducing the number of times you have your hands in the tank.

377 For accurate test results, water salinity must be 35 ppt

Remember that if the water in your reef system is not at natural seawater levels, i.e., a salinity of 35 parts per thousand, then you may get an inaccurate reading when you come to test the water composition.

378 Cleaning the front glass inside and out

Saltwater aquariums, especially reef tanks, have a tendency to develop colonies of algae on the inside of the glass. Much of this will be in the form of soft green algae, but you will also find hard, pink or red, calcareous, coralline algae. Do not bother cleaning the back and sides of the aquarium unless they also act as viewing panels. Bladed scrapers are the best choice for cleaning the glass, as coarse materials can pick up fragments of the hard algae and scratch the front glass. For acrylic tanks, use plastic-bladed scrapers. You can clean the outside of the front glass very efficiently by vigorously rubbing with crumpled up sheets of newspaper.

379 A reef aquarium requires additional maintenance

A reef poses additional maintenance problems over those usually encountered in the aquarium. Calcium can be deposited in such a way as to impair the operation of mechanical equipment. Invertebrates such as snails, sponges and crustaceans can block pipework, leading to flooding and other problems. Clean and service all equipment on a regular basis and do not try to save money by using cheap or old equipment. Equipment failure is a major cause of reef tank disasters.

380 Find time to check on the aquarium every day

Be sure to carry out a daily visual check on the tank. Monitor the behavior, apparent health and number of your livestock. Check the operation of filters, protein skimmers, powerheads, heaters and airstones. Remove large pieces of detritus such as uneaten food or dead livestock.

Tip 380 *Remove dead livestock as soon as possible to avoid pollution problems.*

381 Cover glasses can cause pH levels to drop

Tightly fitting cover glasses, although useful for keeping children and cats out and fish in, can be the cause of pH problems. If you have cover glasses in place, make sure that sufficient gas exchange can still take place, otherwise you may find your pH level dropping.

382 Observe the basic rules of hygiene and common sense

Wash your hands thoroughly, but not with soap or disinfectant, before placing them in the tank. When cleaning equipment, remember to include nets, air lines and siphon tubing. Avoid using colored plastic containers for water changes. Avoid all sources of toxic fumes in the vicinity of the tank, such as furniture polish and cigarette smoke. Avoid sudden shocks to the livestock such as flash photography or children (and adults!) banging on the glass.

Tip 384 *A protein skimmer working efficiently.*

Tip 383 *Rinse out filter media in tank water.*

383 Cleaning filter media in a canister filter

In a fish-only system, keep an eye on the flow rate of water returning from any external filter. If it slows down, the filter needs cleaning. Where there are several layers of foam filter medium in an external canister filter, do not change (or rinse out and re-use) the entire medium at once. Changing a portion at a time (using aquarium water to rinse it out) helps to retain the medium's colony of beneficial nitrifying bacteria.

384 Keep the protein skimmer clean for efficiency

Any buildup of dirt or algae inside the reaction chamber of the protein skimmer (rather than the inside of the collecting cup) may adversely affect the skimmer's efficiency. Fatty substances left on the inside wall will prevent the foam from rising up high enough to be collected, so be sure to keep the protein skimmer clean. In heavily loaded systems this could be necessary once or twice a week.

385 R.O. units need regular servicing too

Do not forget to service your R.O. unit. Replace prefilters as necessary—this may be dictated by the state of your water. Replace carbon at least annually to protect the membrane from bacterial contamination. Replace the membrane at least every five years. Be aware that the life of carbon and membrane will be proportional to the volume of water treated; the greater the volume, the more frequently they should be changed. Manufacturers usually state a capacity that filters and membranes are capable of processing.

386 Not all water test kits are the same

Always follow the manufacturer's instructions carefully when testing water parameters; not all test kits use the same technique, especially when it comes to comparing the colors of the test sample against the supplied color chart. Some tell you to look through the tube; others want you to look down the end from the top. Keep all test kit color comparison charts out of direct sunlight or other bright light to prevent them fading gradually over time and thus giving incorrect results.

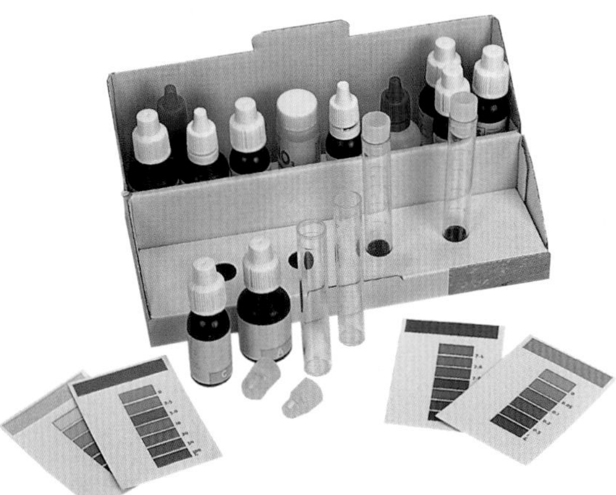

Tip 386 *Follow directions when using test kits.*

Phosphate test

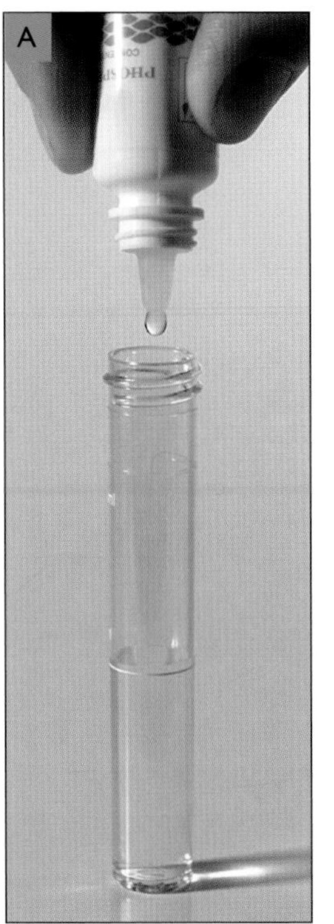

A *Add five drops of the phosphate re-agent to a 1 tsp. (5 ml) sample of aquarium water and gently shake the tube to mix well.*

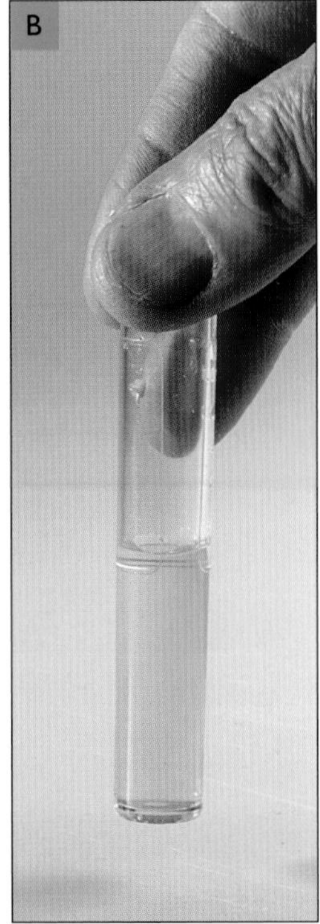

B *Allow the sample to stand for five minutes until the yellow color has fully developed. Check it against the chart provided.*

387 Make the most of your test kits while they are "fresh"

Keep a check on the best before date of the re-agents used in test kits. Liquid re-agents age more quickly than dry ones and this can also produce misleading test results. In view of the short shelf life of some liquid re-agents—some with as short a life as six months—don't skimp on your testing; make sure you get full use out of that test kit!

388 Water tests also tell you what is missing

Don't just test for nasties. As well as testing for undesirable elements, such as ammonia, nitrite, nitrate and copper, test kits also tell you what is missing in respect of good components in the water, such as oxygen, iodine and calcium.

389 Remove salt deposits on metal-halide lamps

The glass screens of metal-halide lamps become obscured by salt deposits, even though the lamps are generally mounted well clear of the water surface. Switch off the lamps and wait until they have cooled down before starting to clean them.

390 How often should I replace my lights?

Change bulbs on a regular basis. If you have a light meter, replace bulbs when the output has dropped by 30 percent. As a rough guide, replace all fluorescent bulbs on a yearly basis, but twice a year for actinic bulbs.

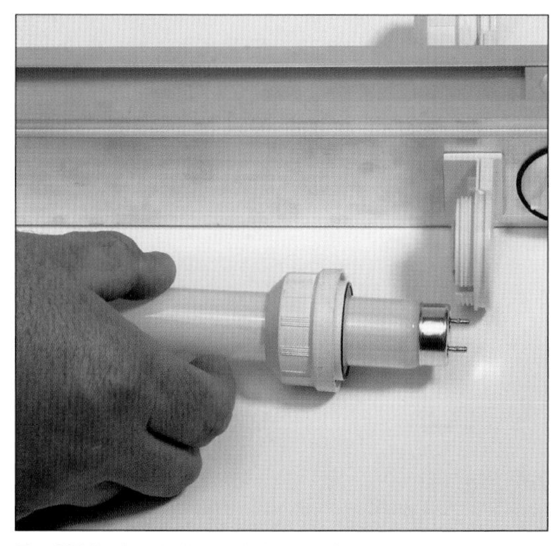

Tip 390 *Replace bulbs regularly, even if they are still working.*

Tip 389 *Clean the glass on metal-halide lamps.*

391 Keeping a logbook has many advantages

Keep a diary or logbook detailing the results of any tests you have carried out, plus a record of your water changes, supplementation and any fish or invertebrates that you have added to the tank. You can of course keep a note of many more developments than the ones mentioned here, such as changes in equipment, servicing details, bulb replacement, etc. The value of keeping records is that it can help you discern any trends in changing water parameters. Also, if you start to run into trouble, your records can help you to identify the cause of your problem.

392 Supply the demand for calcium and carbonate

Initially, water changes will make up for the calcium and carbonate being taken up by the low number of creatures in your reef. However, as the number of animals in the reef increases, the demand for calcium and carbonate will rise. At some point, water changes will no longer be enough to supply these two components.

393 Using and monitoring supplementation

For the smaller tank or the tank with a lower demand, the best choice for supplementation is going to be one of the balanced, two-part additives commercially available. Once you start using a supplement, you will need to monitor the calcium and carbonate levels to make sure everything is working smoothly. When you have established a stable, effective dosing regime, you can cut back on the frequency of testing.

394 Supplementation is not enough in a large tank

With larger tanks, or ones with a high demand for calcium and carbonate, supplements are either unlikely to cope with demand or will become prohibitively expensive to administer. The solution here is to use a calcium reactor or kalkwasser stirrer. These tend to be expensive to purchase but are comparatively cheap to run.

395 Carbonate and calcium levels are inextricably linked

One useful trick of the trade is to monitor just your carbonate levels in the knowledge that if these start to change it is time to check the calcium levels. This works because these two substances are inextricably linked together, so if one changes they both change.

396 Not all salt mixes contain the same level of minerals

If your tests show moderately low readings of calcium and carbonates, one avenue open to you is to use a salt containing higher levels of calcium and carbonate. Some salts may be considered fish salts, having lower levels, while other salts will have enhanced levels of these minerals and can be thought of as reef salts. A change of salt may well keep you running over for some time, but it is probably just delaying the moment until a supplement is needed.

Carbonate hardness (KH) test

A *Counting each drop, add KH re-agent a drop at a time to a 1 tsp. (5 ml) sample of tank water. Swirl the tube to mix.*

B *Initially the sample in the tube turns blue.*

C *As more re-agent is added, the water sample turns yellow. Continue counting the drops added until the yellow color is stable. Each drop added from the beginning of the test represents 1°dH, equivalent to 70 mg per gallon (17.5 mg/L) of carbonate.*

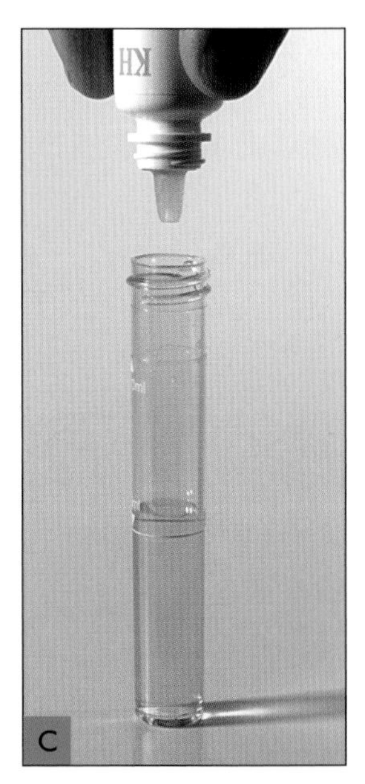

397 How does a calcium reactor work?

A calcium reactor works by passing the tank water through a calcareous medium, while at the same time adding CO_2 (carbon dioxide gas) to the reactor. This process acidifies the water, dissolving calcium and carbonates in a balanced manner, thus making them available for the corals to utilize.

Tip 397 *A calcium reactor in operation.*

398 Do I need to use any other form of supplementation?

Many reefkeepers add supplementary trace elements, such as iodine, strontium and magnesium, along with vitamins. On the whole this is neither necessary nor appropriate. In a general reef aquarium, where water changes are performed regularly, trace element additions are probably not required.

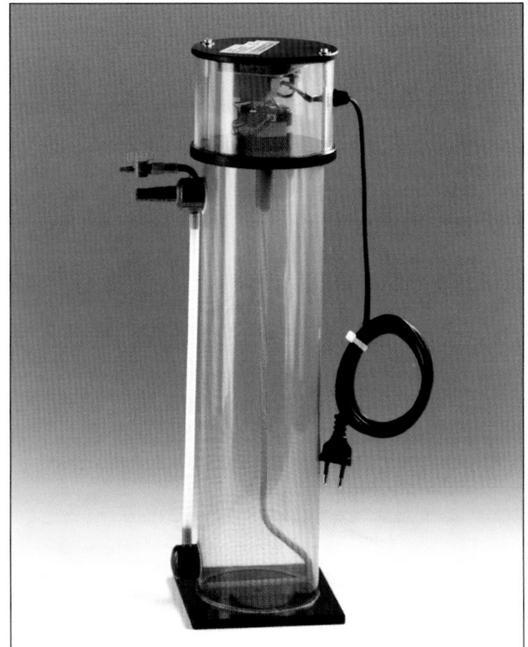

Tip 399 *A kalkwasser stirrer*

What is a kalkwasser stirrer and what does it do?

A kalkwasser stirrer mixes calcium hydroxide into R.O. water, which is then used to top up the tank, thus replenishing calcium at the same time as making up for evaporation losses. Calcium hydroxide has a very high pH. Because of this it is also potentially dangerous to the inhabitants of your tank, children, pets and you. Use it with care.

400 Vitamin supplements should not be necessary

Vitamins should be present in the food you feed to your animals. There is not much point in adding them to the water. If you feed a varied diet, save your money to use on good-quality salt for those regular water changes, instead of adding products of questionable value.

401 Don't expect miracles from aquarium products

There are no miracle products that will automatically give you a perfect reef with healthy corals growing out of the tank. There are no miracle products that will exclude or eradicate all species of pest algae.

402 Should I add iodine to my reef aquarium?

The dosing of both iodine and strontium is somewhat controversial. There is conflicting scientific evidence as far as both these minor elements are concerned—and interpretations of the evidence also differ. With regard to iodine, there is a great deal of anecdotal evidence to support its addition. If you want to add it, do so with food so that it can be utilized directly by the animals. Do not exceed natural seawater levels of iodine—about 0.24 mg per gallon (0.06 mg/L).

403 Will my aquarium benefit from adding strontium?

Strontium is probably best ignored, as it is incredibly difficult to get any sort of meaningful result from the test kits available to aquarists. Unless you are an advanced aquarist, the corals you are keeping are unlikely to benefit from supplementation. One often-repeated axiom of reefkeeping is that if you can't measure it with a test kit, then don't add it!

404 Check magnesium levels before supplementing them

Often, magnesium additions only seem to be needed if inappropriate methods of adding calcium and carbonates are being employed. But if something does not seem right in the tank and all other parameters check out, it would be worth checking magnesium levels and then supplementing them if required.

Tip 401 *Careful monitoring of the water conditions helps to keep this superb tank looking its best.*

405 Macro-algae have a place in the saltwater aquarium

A healthy growth of macro-algae (small, leafy, plant-type growth as opposed to unicellular algae, which produce the "green water" effect) is not necessarily a bad thing, if you can keep it in check. Herbivorous fish will do this for you, but you may have difficulty in establishing just the right ratio of fish to algae to maintain the desired visual effect.

Tip 407 *A strong growth of* Caulerpa *spp. macro-algae.*

406 Cleaning rocks the natural way, one by one

A natural way to clean algae-covered rocks is to remove them (one at a time for convenience) and place them in a separate tank with some herbivorous fish. Once cleaned, the rock can be exchanged with the next algae-covered one from the main tank.

Tip 405 *A harmonious mix of* Caulerpa*, disc anemones and fish.*

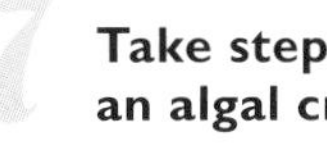

407 Take steps to avoid an algal crash

Without herbivorous fish, regular manual pruning will keep the algae growth under control, but be on the lookout for a sudden algae crash. A mass death of algae can reduce oxygen levels very quickly and only a massive water change will deal with the situation.

408 How to establish stability with macro-algae

With 24 hour a day illumination of a mud system or deep sand bed (DSB), the risk of an algae crash due to the sexual reproduction cycle of *Caulerpa* becomes negligible. At the same time, this system can help to reduce fluctuations in pH over the course of the day.

409 Your tank can become overgrown just like a pond

Algae growth can be a problem if you haven't got any herbivorous fish in the tank. Just as a pond can suffer from too much plant life, so too can the saltwater aquarium. This is particularly the case during the hours of darkness, when photosynthesizing action by algae not only stops but might be regarded as going into reverse, with oxygen being taken from the water rather than excess oxygen being released into it. Keep the water circulation going during the night, add extra aeration if the weather gets very warm, and prune out some of the greenery.

Tip 409 Caulerpa *spp. macro-algae and cardinal cleaner shrimp.*

410 Macro-algae absorb unwanted nutrients

When employing macro-algae in an algal filter, such as a mud system or deep sand bed (DSB), you will get better results running with higher levels of illumination above the algae, and higher flow rates through the filter. Harvest the algae regularly as a means of removing unwanted nutrients from the system.

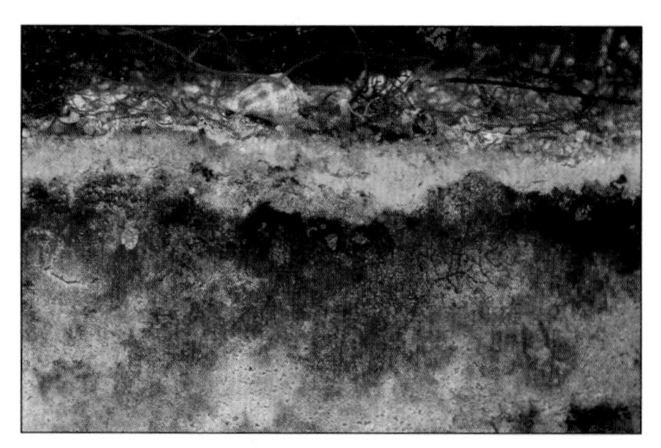

Tip 411 *A deep sand bed filter with macro-algae.*

411 A wide-ranging strategy for dealing with pest algae

Pest algae are always a potential problem in the reef aquarium given the amount of light needed to sustain corals. Light, along with phosphates and nitrates, can be viewed as fertilizer for undesirable algal species; reefs need the light, so concentrate on limiting phosphates and nitrates. Live rock, deep sand bed (DSB) and mud systems will all successfully process nitrate: use a phosphate-removing medium when first setting up the system, keep an eye on phosphate levels with the aid of a test kit, and then renew the medium whenever you see signs of phosphate levels increasing. Also, employ invertebrate herbivores as much as possible. If there were no herbivores on the reefs there wouldn't be any reefs!

412 *Caulerpa* is a vigorous grower—keep it in check

The main genus for providing macro-algae is *Caulerpa*. It comes in many forms, with leaves of varying shapes in shades of green from blue-green to emerald. Do be wary when growing macro-algae in the display portion of a reef system, as many species of *Caulerpa* can become invasive, growing over certain species of corals, depriving them of light and killing them in the process.

Tip 412 Caulerpa racemosa *has a distinctive leaf shape.*

413 Raise macro-algae in a dedicated refugium

Usually encouraged by high nitrate or phosphate levels, algae can be grown deliberately in a refugium (a separate, brightly lit mini-tank sharing the main water flow) to extract these compounds from the water. Feed any harvested algae to your herbivorous fish; otherwise, put it in the compost bin.

414 Moving on to a larger tank—first thoughts

The day will come when you decide it is time to move up to a larger aquarium. Transfering the contents of an existing reef to a new, larger tank may seem daunting, but it is really just a matter of good organization. Before conducting the transfer, ask yourself a couple of questions. What is the difference in volume between the two tanks? Is the new tank going to be positioned in the same place as the existing one?

415 Establish the volume of water in the tank

First, you need to ensure that you can top up the difference in water volume when you transfer the contents of your existing reef into the larger tank. This is not as important if your new aquarium does not have a sump, as you can then just run with a lower water level for a time. If the new tank does have a sump, then you will need the correct volume of water for the system to run.

416 Think ahead—save water from water changes

When transferring an existing reef to a larger tank, it helps to think of the operation as being a larger-than-usual water change. It might be inadvisable to add more than the existing volume in new saltwater in one procedure; this would be the same as a 50 percent water change. Some aquarists plan ahead when upgrading to a larger tank by saving water from their water changes to reuse when filling the new tank.

417 An intermediate water container is essential

As you need to dismantle the reef from the top down, and then rebuild from the base upward, you will need to employ a holding tank of some sort. This could be a third aquarium, or a large, food-safe plastic container, such as a household coldwater tank or a koi vat. You may be able to borrow a suitable container from your local aquatic store.

Tip 417 *A koi vat*

Moving aquariums

418 Many aquarists start off their first reef with a modest-sized aquarium. However, with increasing experience, ambition and finances, the day will come when it is necessary to change to a larger aquarium.

1 Siphon some water from your existing reef aquarium to the holding tank—just enough to cover the corals for the short time they will be there. Move the corals to the holding tank. While you are doing this, water can continue siphoning or pumping into the new tank.

2 It may help you to arrange the material in the holding tank in three "zones": corals, live rock for building the reef infrastructure and miscellaneous rock for decorative purpose. Once the corals are in the holding tank, move the upper layers of live rock into the same tank.

3 Cease siphoning the water into the new aquarium when you have reached the minimum depth required for any fish present in the existing tank. As you move the corals and live rock into the holding aquarium, also transfer any mobile invertebrates you encounter.

4 Place the lower levels of live rock directly into the new aquarium, followed by the upper layers of rock, then the corals and decorative live rock from the holding tank, and lastly the mobile invertebrates. Agitate the rock to remove trapped air.

5 Net the fish and move them to the new tank, being careful not to stir up detritus from the bottom of the tank. This would obscure the water and place additional pressure on the fish at an already stressful time. Before moving the substrate over to the new aquarium, it is worth rinsing it through with aquarium water to limit the amount of detritus you transfer.

Tip 418 *Moving a large reef aquarium is a complex process.*

419 Get the water moving in the new aquarium

If possible, start up the pumps. The water movement will aid gaseous exchange and help wash away mucus from the corals. Turn on heaters, as the temperature will probably have dropped during the changeover. If the new aquarium employs a sump, you may not be able to provide water flow and heating until you have topped up the system to the correct volume of water.

Tip 420 *Do not leave any creatures behind in the old tank.*

420 Be sure to transfer all the inhabitants of the tank

Once everything is in position, slowly top up your new reef with any additional water that is required. Check the old tank one last time to make sure you haven't left any animals behind, particularly creatures that live in the sand, and snails that may be attached beneath the aquarium's strengthening bars.

421 Transferring to a new tank in the same position

If a new aquarium is to occupy the same position as the existing one, follow the procedure described, but in addition you will need to transfer all the rockwork to the holding tank and bag up all the fish (in clear water, not water containing suspended detritus). Drain as much water as you can before moving the existing tank, although it may be possible to leave the substrate in place with just a bare covering of water. Move the new aquarium into position then continue as before.

422 Moving your aquarium when you move house

The procedures for upgrading a reef tank can also be broadly applied when moving house, i.e., the order of breaking down the tank and putting it back together. However, instead of using a holding tank, bag up the fish, invertebrates and rockwork. Ask your local aquatic shop if they can supply you with several fish bags and boxes. The Styrofoam boxes used for delivering fish to the store are exactly what you need for transferring your livestock to the relocated aquarium.

Tip 422 *Transfer fish in plastic bags in a Styrofoam box.*

Better Ongoing Care

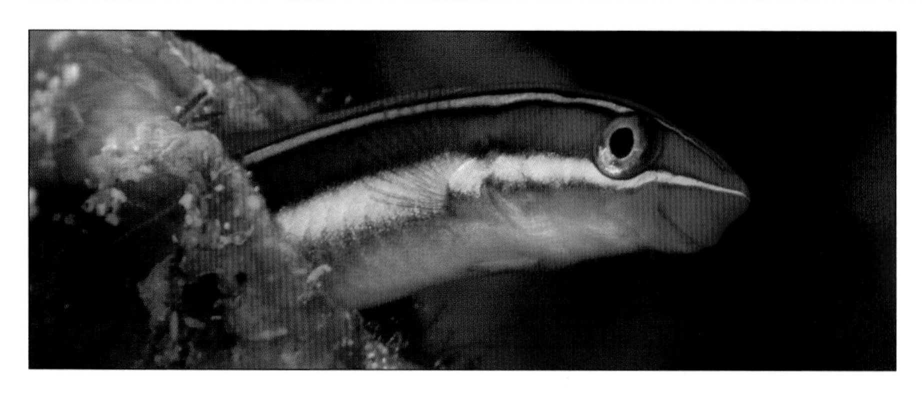

Tip 423 *Do not feed the fish for a day before the move to avoid the risk of polluting the transportation water.*

423 No feeding before the day of the move

Do not feed the fish for a full 24 hours before you move your aquarium into your new home. If you do, they will run the risk of being poisoned by their own waste during transportation in the close confines of their containers.

Photographs are an excellent aid to recreating an aquarium display.

Tip 424 *Keep a photographic record of the old tank setup.*

424 Take photos to remind you how the setup looked

If you intend to re-create your reef as it stands, take plenty of reference photographs from all angles before you strip it down. This will make it much easier to build up the reef again, rather than relying on your memory. Label the boxes to indicate whether they contain coral, base rock, etc., to speed up the rebuilding process.

425 Don't move house and tanks on the same day

Moving an aquarium on the same day as moving house is virtually impossible. If you can, arrange to move the tank in advance. If your vendor will not agree to this, consider moving the tank to a temporary location at a friend's or family member's house.

426 Feed carefully just after you have moved the aquarium

After the move, be conservative when feeding your reef, and keep a close eye on the various water parameters. The reef will have gone through a great deal of disruption, and it is almost inevitable that some smaller, hidden animals will have died. As a result, you could well experience some minor pollution episodes. Resume your regular pattern of water changes as soon as possible.

427 Electrical safety is paramount at all times

Before moving or upgrading a tank, remember to switch off and remove all electrical equipment. Do not risk dropping heavy rockwork against heaters in the tank. Rig up additional external lighting to work by if you are in a poorly lit area. Electricity is always an issue—and not just when you are moving tanks. To make things safer, all electrical equipment should be covered by a grounded circuit-breaker. If you do not have one on the outlet, or at the fusebox, buy a plug-in adapter.

428 Take as much original water as you can manage

Arrange well in advance to beg, borrow or buy as many water-carrying containers as you can. To avoid subjecting your fish to a massive water change, aim to transport as much of the existing tank water as possible. When you are installing the tank in your new home, fill it with the old transported water and top up with new saltwater. If you have not been able to make up new saltwater in advance you will have to make it up on the spot using tap water, fierce aeration and a tap water conditioner.

Tip 428 *Moving a reef aquarium is possible if you plan well ahead.*

Better Breeding

429 Are you prepared for this labor of love?

Breeding saltwater fish is a labor of love. Invariably it will lead to sleepless nights, incredible failures and significant expenditure. Few people have ever made money out of selling saltwater fish. However, it is also one of the most rewarding aspects of the hobby and every aquarist should attempt it at least once.

430 Breeding saltwater fish is difficult but not impossible

Most saltwater fish lay eggs that hatch into larvae, whose first instinct is to swim into the plankton layers found at the ocean's surface at night. Here they feed on microscopic animal life until they undergo metamorphosis and resemble the fish we know. For this reason, many species will be difficult if not impossible to breed, but this should not deter you from trying.

431 The basic requirements for successful breeding

To encourage any aquatic animal to breed, you must first provide good water quality, followed by a compatible mate, a safe environment and good nutrition. Some species may have further requirements, such as cues that stimulate breeding in the way of, perhaps, the lunar cycle or changes in salinity or temperature, plus other variables. Species that need these additional cues are better left to the experienced breeder.

432 Remove adult fish to a spawning tank

Many saltwater fish will happily spawn in a reef aquarium. If you are serious about raising the larvae, you may have to remove the adult breeding pair to a separate aquarium. Here you can observe the potential parents and, by providing a limited number of spawning sites, monitor egg development more easily too.

433 Which species are likely to breed in the aquarium?

The most commonly raised saltwater fish species are Banggai cardinals *(Pterapogon kauderni)* and clownfish—usually *Amphiprion ocellaris*, or western clowns—although most species of clowns are relatively easy to raise. Damselfish species, such as gold chromis, neon gobies *(Elacatinus evelynae* and *Elacatinus oceanops)* and various dottyback species also breed readily.

434 The Banggai cardinal is a rewarding first fish to breed

The easiest of the commonly available species to breed is the Banggai cardinalfish *(Pterapogon kauderni)*. In fact, with this species, other than making sure you have a pair, feeding them well and giving the young a safe, predator-free environment, you will have very little to do! In this species, as with most other cardinalfish, the female lays a ball of eggs that the male takes into his mouth for incubation. What separates the Banggai from other cardinalfish is that the young are highly developed when they finally emerge from the male's mouth. They are large enough to accept newly hatched brine shrimp and frozen foods of appropriate size, so raising them to maturity is relatively straightforward.

Tip 434 *Adult Banggai cardinalfish, with juvenile (inset).*

***Tip 433** Western clownfish* (Amphiprion ocellaris) *are straightforward to breed.*

435 It is not possible to sex juvenile Banggai cardinals

Banggai cardinals are of fixed sex, which can cause problems when you want to establish a pair, as male and female juveniles are indistinguishable. However, once the male becomes sexually mature he will be intolerant of other males, and fierce, potentially lethal fighting will ensue. Form pairs by putting half a dozen juveniles together in a tank and look out for two fish associating closely together. Be ready to remove the remaining fish to safety.

436 Many reef fish change sex, some more than once!

Certain species of reef fish have evolved a method of sex determinism. Instead of being born a fixed gender, many of these fish start off being of indeterminate gender, developing as males or females in response to various stimuli. Some species are capable of then changing sex from male to female, while other species even have the ability to revert to their original sex after the initial change.

437 Sex determinism can help the saltwater aquarist

Clownfish are a fine example of how this sex change can work for the aquarist. At its simplest, just put two juvenile clowns together and in time you will have a pair. It is best to refine this slightly by choosing juveniles of unequal size. The larger fish will become the dominant of the two and become female, while the smaller fish will become male. By choosing fish of different sizes, you reduce the risk of possible injury or even death that can occur when two fish of equal size are battling for dominance.

438 Sex changes occur within a group of clowns

A hierarchy exists within a group of clowns, with the dominant female at the top, followed by the male—her mate. All the other clowns in the group remain of indeterminate sex. If the female dies, then the male, being the dominant fish, will become female, and one of the remaining "no-sex" clowns, probably the largest, will become a male.

439 Give clowns a removable spawning substrate

Western clowns *(Amphiprion ocellaris)* are the most widely raised species. As with other demersal spawners, they lay their eggs on a clean, flat surface, so it is relatively easy to retrieve the eggs if these are attached to a removable piece of substrate, such as a tile, flowerpot or length of plastic pipe. Remember to use food-safe materials for spawning. You can then transfer the eggs to a separate rearing tank and attempt to raise the larvae in a predator-free environment.

440 Some fish are fussy about their spawning site

Some saltwater fish species are very particular about the area of the tank in which they will spawn, and often rearrange it. If you wish these fish to breed, be prepared for some disruption. One example is the female maroon clownfish *(Premnas biaculeatus)*, which often refuses to have any rocks or sand near her spawning site and will clear these accordingly, often smothering or damaging sessile invertebrates in the process.

441 Acquiring a pair of dottybacks is not difficult

In order to achieve a pair of most of the commonly kept species of dottyback originating from the Red Sea, you need only introduce two juvenile specimens into the aquarium. These will differentiate into a male and female specimen that should spawn regularly, even in a very small aquarium. Examples of such species include the orchid dottyback *(Pseudochromis fridmani)*, sunrise or blue dottyback *(P. flavivertex)* and the orange or neon dottyback *(P. aldabraensis)*.

442 Male dottybacks "disappear" while guarding eggs

When the male dottyback is guarding his ball of eggs, he will disappear for a number of days—much to the alarm of the vigilant aquarist. The female feeds readily during this time in order to gain the energy and resources she needs for the next breeding cycle. An established pair of dottybacks might breed as often as every nine to ten days, even in the smallest aquarium.

Clownfish spawning

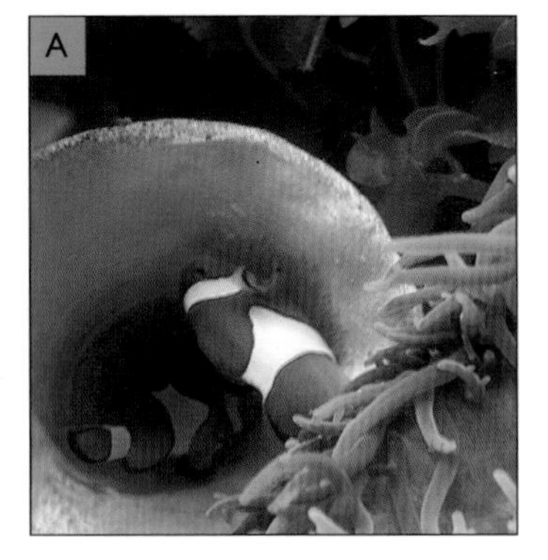

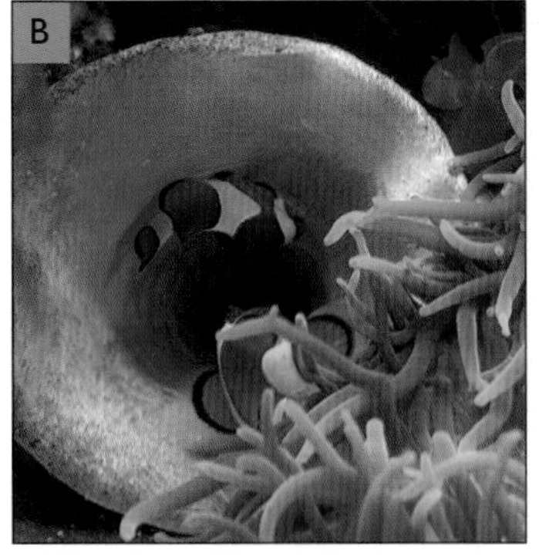

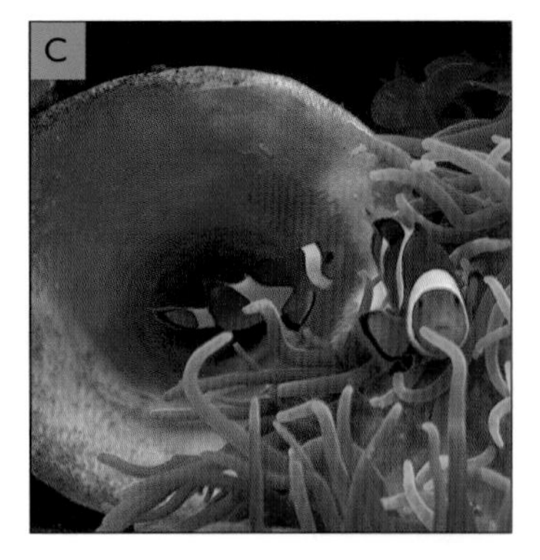

A *Having selected a suitable spawning site (here a clay pot), the fish first clean the site and then start taking turns at spawning runs.*

B *The female, the larger fish, lays her eggs and is then followed by the smaller male, who fertilizes them with his sperm.*

C *Once the eggs have been laid and fertilized, the male looks after the eggs, while the female takes on the duty of guarding the site.*

443 A pair of twinspot gobies make fascinating subjects

Where fish are found in pairs in their natural environment, you should try to keep them like this. Inevitably, they will be much more contented and in return should reward you with natural behavior. A pair of twinspot gobies *(Signigobius biocellatus)* that are feeding well will keep you enthralled for hours at a time, and if they breed, you will be captivated for weeks.

444 Don't expect neon gobies to spawn continuously

Neon gobies *(Elacatinus oceanops)* are also relatively easy demersal spawners, but unlike clowns, they can be more seasonal or have brief periods of spawning activity interspersed with long periods of inactivity. There is also evidence that they pair up seasonally, so it may be wrong to expect them to reproduce continuously with the same partners.

445 Breeding seahorses for other hobbyists

If you have established an aquarium for one or two pairs of seahorses, you can be fairly certain that they will breed. Although it is tempting to try to rear every brood that emerges from the male's pouch, you have to be practical. If you cannot set up a breeding program that will handle large quantities of young, then try raising a few for the experience and to give something back to the hobby. Any seahorses offered for sale should be captive-bred, and by growing some to tradeable size you help to ensure that there are plenty available in the hobby.

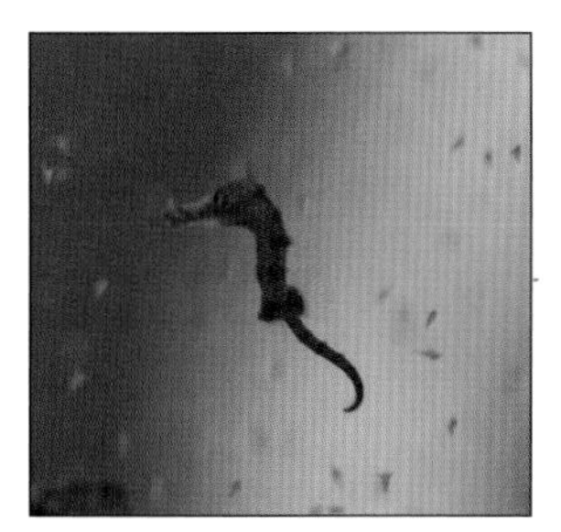

***Tip 445** Young seahorses in food-rich water.*

446 Breeding success comes with experience

When fish start to spawn, especially if they are young specimens, their first efforts may fail. To produce a batch of fertile eggs successfully, the female must learn to lay her eggs in a tight pattern, making it easier for the male to swim across and fertilize them. The male must learn how to deliver his sperm efficiently to the eggs. Observe your fish during this learning period and record how often they spawn and where, and if there is any stimulus that initiates spawning. This will help you to predict when they are going to spawn—useful information that allows you to ensure that the prospective parents get optimum feeding when they need it and plan food-raising schedules for the anticipated larvae.

Tip 446 Centropyge resplendens *male stimulating female.*

447 Nutrition is the key to raising fish larvae

Spawning is easy; the real difficulties start when you come to raise the larvae. Most problems stem from nutrition, as the larvae require a combination of live foods in the form of phytoplankton and zooplankton. Producing and maintaining these live food cultures is the key to the entire process of raising saltwater species. If you cannot provide these essential live foods in sufficient quantities, it is very unlikely that you will manage to raise any larvae to adulthood.

448 Prepare larval food cultures well in advance

Be sure to have food cultures up and running before there are any larvae to raise. You will lose all the larvae, unless a friendly breeder takes pity on you and helps you out with cultures of their own. Phytoplankton is the base of the food chain when raising saltwater larvae. After phytoplankton come rotifers, followed by the newly hatched first larval stage (nauplii) of brine shrimp.

449 Rotifers—a vital link in the food chain

Saltwater fish larvae require easily digestible, small particle-sized food—far smaller than brine shrimp nauplii—and it must be moving, so they are attracted to it. Rotifers reproduce themselves quickly, making them ideal for this purpose. By feeding the rotifers on phytoplankton to keep them alive and reproducing, you ensure that it is the phytoplankton that supplies most of the nutrition to the larvae. As long as the water in the rotifer culture is kept green, there is food available to them. If the water clears, add more phytoplankton until the culture has regained its green tinge. Different species of phytoplankton can be fed to the rotifers depending on the fish species being bred.

450 Growing larvae need larger foods than rotifers

As the larval fish grow, there comes a point when they can be expending as much energy chasing rotifers as they are actually gaining from eating them. At this point, switch to newly hatched brine shrimp nauplii, less than 12 hours old. Introducing brine shrimp does present a risk (see Tip 464), so be sure to limit the amount introduced into the tank and observe the larvae to make sure there are no problems.

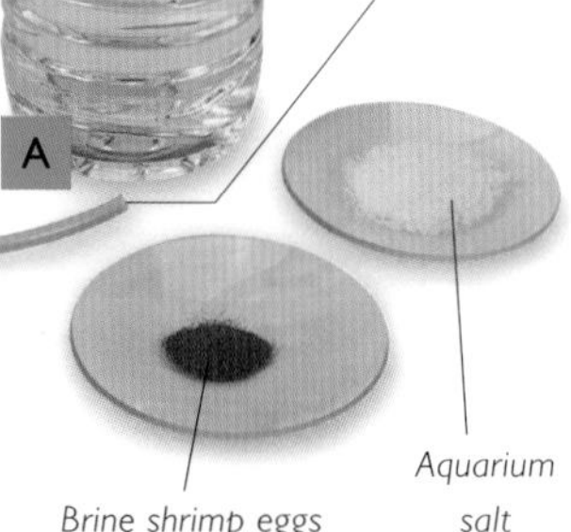

A *Half-fill a 32-oz (1 L) plastic bottle with fresh tap water and add 1½ teaspoons of sea salt and ¼ teaspoon of eggs.*

Drop in an air line connected to an air pump for circulation.

Brine shrimp eggs

Aquarium salt

B *Keep the eggs in constant motion with some air. Rigid air line is best for this, but flexible tubing will also work if you position it properly. Remove the air line after 36 hours and let the water stand for 30 minutes.*

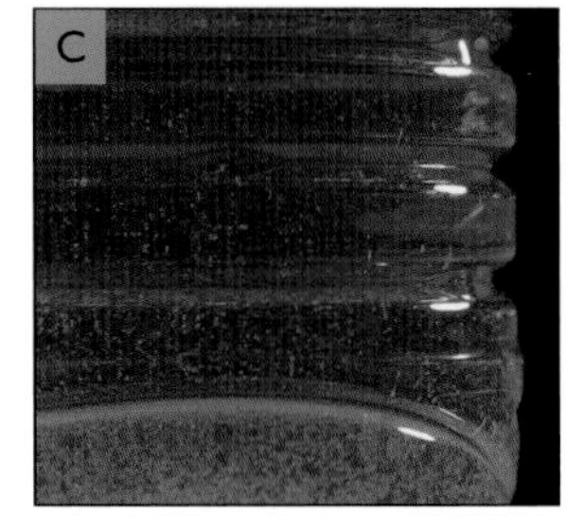

C *Separate the shrimp from the salty water by pouring them through a funnel lined with a paper towel. Rinse them with fresh water and feed them to the larvae.*

451 Feeding dry foods makes life easier for the breeder

The aim of many breeders is to wean the larvae onto dry foods as early on as possible. This can help limit brine-shrimp-associated losses, but more importantly it makes life easier for you, the breeder. It allows you to divert your attention away from raising the various live foods and to concentrate more on maintaining water quality in the raising tank.

452 Preparing the larval raising tank and live food

When you have enough information about the spawning pattern of your fish and feel you are ready to try raising a batch of fry, set up live food production and the larval-raising tank. Time your setup to coincide with a spawning so that the tank does not lie empty too long, but remember you must have that all-important live food ready in sufficient quantities to have any chance of success.

453 A dedicated larval tank is essential

To raise larvae you must set up a dedicated larval tank. With the exception of Banggai cardinals (and certain species more suited to the advanced aquarist), all the commonly raised species initially require an open water (pelagic) environment. This is why it is virtually impossible to raise larvae in a display tank; the larvae need to remain suspended in the water column, swimming in food, safe from predation and safe from being sucked into pumps and filters.

454 Setting up a basic larval-raising tank

Requirements for different species may vary slightly, but a useful general-purpose setup, suitable for clowns and shrimp, would be: a 24 x 12 x 12 inch (60 x 30 x 30 cm) aquarium, heater to suit, air pump, length of flexible air line, short length of rigid air line, thermometer, ammonia alert badge and some material to shade the tank from light. Set up the tank on a piece of white Styrofoam. This will enable you to see waste material on the bottom of the tank, making it easier for you to siphon out detritus when cleaning. Cover the back and sides of the tank with black material, and prepare a removable black cover for the front glass. The reason for blacking out the tank is to make it easier for the larvae to see their prey and to prevent them from being distracted by light from outside the tank, which could lead to them crowding together in one spot rather than spreading evenly throughout the tank. When the larvae's eyes have developed a little more, five to seven days depending on the species, you can discard the front cover.

A larval-raising tank

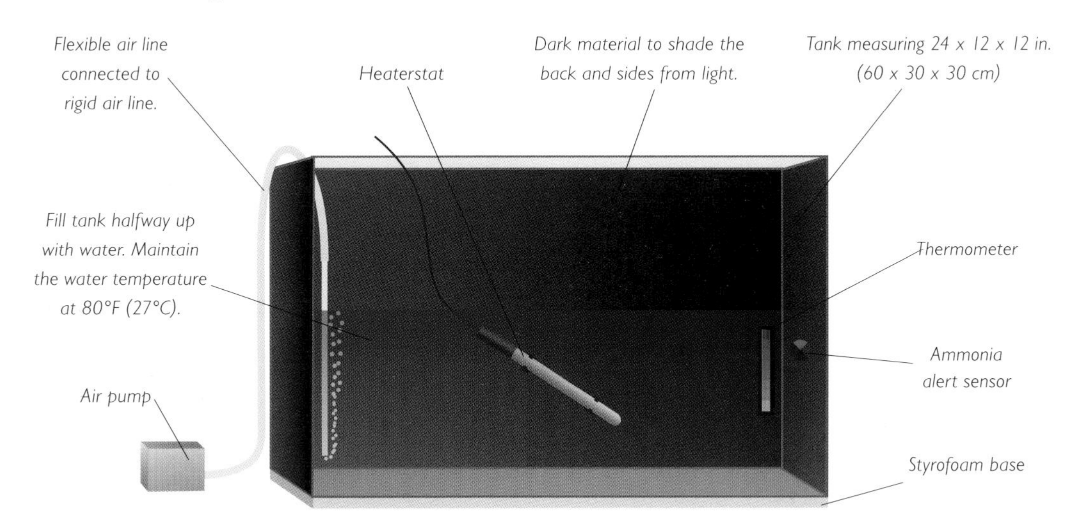

455 Filling the larval tank and regulating the air flow

Fill the tank about half full with water from the parental system, fix the rigid air line into a convenient corner of the tank and start the air pump. Regulate the airflow to provide good water circulation throughout the tank, but without excessively vigorous bubbling, which could batter the larvae. If using a heaterstat, turn it on, ensure that it is covered by the water and adjust it over a period of 24 hours to maintain a stable temperature of 80°F (27°C).

456 Why is the larval tank only filled halfway?

Starting off with the larval-raising tank half full has two main benefits. First, by minimizing the volume of water, you can more easily maximize rotifer density and thus make it easier for newly hatched larvae to hunt them down. Second, you can avoid making water changes that involve siphoning while the larvae are really tiny; instead, gradually add water from the parental tank to dilute pollution. Top up the parental tank with new saltwater to make up the shortfall.

457 Keep the breeding tank free from obstructions

Limit the amount of hardware in the larval-raising tank. Any associated small gaps around sucker clips, for example, represent potential death traps for larvae. Many larvae are incapable of maneuvering themselves out of tight spots and can become trapped and die. Considering the numbers of larvae you will have in the tank, any losses will be insignificant, but it does represent a source of pollution in an unfiltered system that you need to be aware of. Always siphon out dead larvae to minimize pollution. If the tank is in a position where the temperature remains stable and no additional heating is needed, it is an advantage not to include a heater.

458 How long should the lights remain on over the tank?

Provide a regular photoperiod of 12 to 16 hours for larval-raising tanks, using standard T8 fluorescent lighting. In times of low rotifer production, you can use a longer night period to enable the rotifers to catch up. The longer the day, the more food the larvae will eat, so by lengthening the night they will consume fewer rotifers, leaving more rotifers to breed to replenish their numbers.

A pair of adult gold chromis ready to breed.

459 What specific gravity works best in a larval tank?

One thing to consider is the question of what specific gravity to use when embarking on a breeding project. It is accepted practice to maintain salinity at natural seawater levels of 35 parts per thousand (ppt) equivalent to a specific gravity (S.G.) of 1.026 at 79°F (26°C). When raising larvae and their associated live foods, it becomes advantageous to work at lower salinities. The reasoning behind this is that rotifers grow and reproduce at a faster rate when kept at lower salinities than that usually used to maintain saltwater animals. Rotifers are intolerant of a change in specific gravity greater than seven points—i.e., a difference of 0.007. They grow best in a specific gravity of 1.007–1.014. Thus, we can use a rotifer culture S.G. of 1.014 and a larval tank S.G. of 1.021. As a result of all this, some breeders maintain their brood stock at a lower than normal S.G.

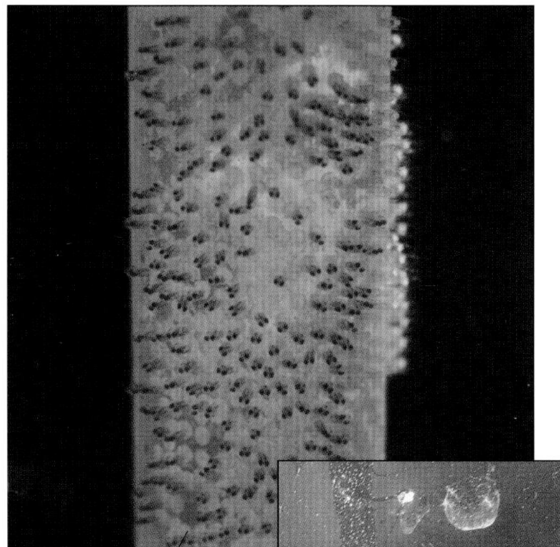

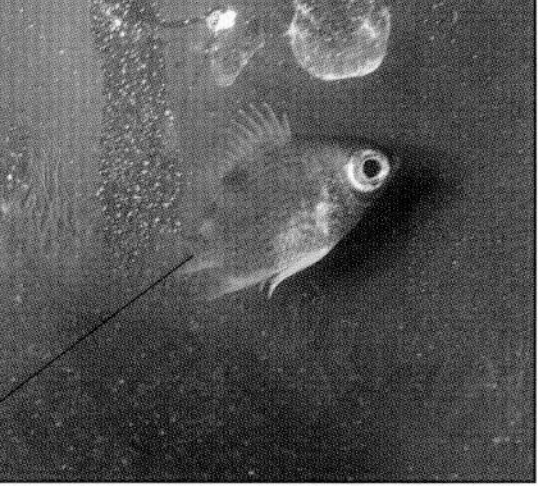

Tip 461 *Two-day-old gold chromis eggs laid on a plastic pipe.*

Ten-week-old juvenile.

460 Use air line tubing to siphon out dead larvae

When siphoning food or dead larvae from the tank, you will have greater control if you use flexible air line tubing, ideally connected to a length of rigid tube. This will reduce the risk of siphoning out healthy larvae.

461 Preparing for the fertilized eggs to hatch

The evening before hatching is due, transfer the substrate holding the eggs to the larval tank. Position it adjacent to the rigid air line so that bubbles are gently buffeting the eggs, simulating the fanning of the parent fish. Ensure the eggs are being kept constantly in motion—more is better than less. Then add enough live phytoplankton to color the water and a small starter culture of live rotifers.

462 Feeding the newly hatched larvae with rotifers

Next morning you should have a tank full of tiny larvae. This is when the work starts. Add rotifers to the tank, aiming for a density that means the larvae need not swim more than their own length before encountering a rotifer. Also increase the phytoplankton density at this time; it will do double duty as both food for the rotifers and, by utilizing some of the waste products from the larvae, will also help to maintain water quality.

463 The early days in the life of the larvae are critical

Over the next five days or so keep up the rotifer density and keep a green tinge to the water with additions of phytoplankton. These first few days are where it often all goes wrong. Losses experienced at this time are usually caused by lack of food; if you cannot maintain rotifer density, the larvae will starve.

464 How long should I continue to feed larvae on rotifers?

Somewhere between day five and day ten, depending on the species, you can move on to newly hatched brine shrimp nauplii. Not all larvae will grow at the same rate, so continue with rotifers at the same time. The introduction of brine shrimp presents a risk, however; the larvae, after days of gorging on rotifers, will try the same with brine shrimp and this can lead to death by choking or overeating. Limit the feeding of nauplii to about 25 per larva per day, spread over the course of the day.

465 When should I start to introduce dry foods?

Many breeders try to keep the nauplii-feeding stage as short as possible to limit losses. To do this, you need to introduce dry foods early on. Start with powdered flake and progressively increase the food particle size as required. You can take a good-quality flake food and just crumble it between your fingertips to a size suitable for your young fish.

466 Encourage young fish to accept dry foods

The young fish will need to learn that this strange, unmoving stuff on the surface of the water is edible. Try feeding new foods first thing in the morning, when the larvae are hungry. They will accept dry food gradually, probably following the example of one of their more adventurous brethren. The downside of moving on to dry foods is pollution. Clean out all uneaten food and minimize the amount you introduce until you are sure it is being eaten.

467 Keep a watchful eye on ammonia and pH levels

Use an ammonia alert sensor to warn you of any problems with water quality. Do not be unduly worried about low pH readings on larval-raising tanks. At a pH of 7.5, not unheard of in a larval tank, you have the advantage that any ammonia present is nontoxic. In an unfiltered tank, ammonia is almost inevitable, but as pH increases so does ammonia toxicity, so this is one occasion in the saltwater hobby when a low pH is a good thing.

468 When do I need to start filtering the tank?

Next fill the tank to its maximum operating capacity and start making small daily water changes, again with water from the parental tank. When you have weaned the youngsters onto a dry food diet, add an air-powered sponge filter to their tank. It is a good idea to run this in the parental tank first to build up beneficial bacteria in the sponge media.

469 Look out for signs of aggression among fry

Once weaned onto dry food, it should just be a matter of keeping the young fish adequately fed while keeping the water quality high. Depending on species, however, you may have to find ways of dealing with aggression. Some species may need segregation, while others may need to be kept more densely to allow any aggression to be shared equally among them.

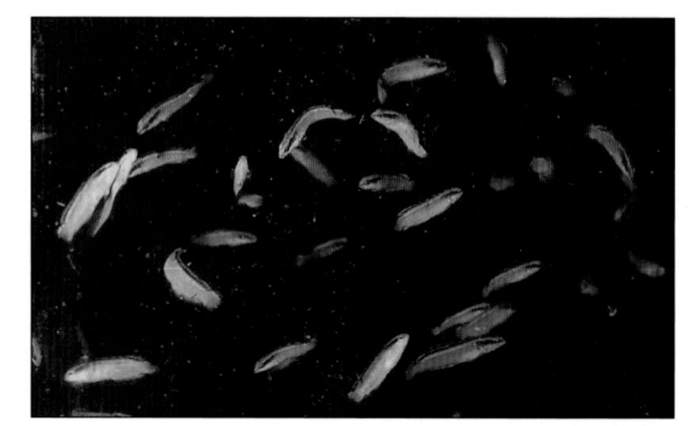

Tip 469 *A tank full of healthy young fish is a just reward for hard work.*

470 Asexual reproduction in anemones and corals

Many invertebrates will reproduce readily in the saltwater aquarium, although most will do so by asexual means. The most common form of asexual reproduction in the home aquarium is through a process called pedal fission, whereby the foot of an anemone or polyp splits, producing two daughter colonies from the original parent colony. Also, baby colonies form directly on the stem of soft corals such as *Sarcophyton* spp.

Tip 470 *Baby colony budding from stem of* Sarcophyton *spp.*

471 *Lysmata* shrimp are not difficult to breed

Shrimp of the genus *Lysmata*, commonly referred to as cleaner shrimp, are probably the easiest invertebrate species that you will set out to breed. All you need are two shrimp of the same species and you have a pair. Both will carry eggs on a regular basis, usually releasing larvae late at night. *Lysmata wurdemanni* (peppermint cleaner shrimp) are by far the easiest to breed because they are simple to feed. The setup described earlier will work for raising the young, but pay extra attention to cleanliness in the larval tank and prevent any buildup of hair algae—the tiny shrimp can become caught up in the algal strands and die.

472 More about peppermint cleaner shrimp breeding

Peppermint cleaner shrimp are voracious feeders and can be raised on dry foods from day one. If underfed, they will resort to cannibalism and eat their siblings. The problem with feeding dry food is the tendency to promote hair algae. As soon as the larvae can manage them, offer adult brine shrimp and increase phytoplankton usage. One brine shrimp can keep a two- to three-week-old larva busy for several hours.

473 Developing crustaceans go through molts

Shrimp and other crustaceans go through a number of molts as they grow. They can gain or lose appendages, giving them a different appearance every few days. Each time they molt, there is a risk of losses. The better their nutrition, the less risk of losses during molts.

474 After the final molt the shrimp will settle down

All the time that shrimp are growing and molting, they remain free-swimming. When the final molt takes place, do not be fooled into thinking that you have lost them all. The larvae settle in and become true juvenile shrimp, at which point they will become substrate-orientated, only venturing into the water column when food is introduced.

Tip 474 *Juvenile and adult* Lysmata seticaudata.

Better Healthcare

475 Stable, stress-free conditions are the key to good health

One of the best ways to ensure the good health of your charges, both fish and invertebrates, is to maintain good-quality, stable water conditions and a stress-free environment. Most health problems usually come down to environmental problems, poor choice of species and equipment failure or malfunction, or aquarist-induced problems, such as inattention, lack of knowledge and "accidents."

476 Prevention is always better than cure

One obvious way to prevent disease is not to risk adding it to the tank in the first place. To achieve this, fish should be quarantined for the requisite length of time (usually 14–21 days), thereby ensuring that a new addition is perfectly healthy before placing it in the display tank. Bear in mind that invertebrates, water from dealers' tanks or apparently healthy fish can all be responsible for bringing disease organisms and parasites into the display aquarium.

477 Look out for changes in fish behavior

Make a point of closely observing your fish on a regular basis, as the first signs of disease are often seen as behavioral changes. The fish may take to sulking in a corner of the tank, may generally look "off color" or may actually refuse to feed. However, do not become obsessed with the slightest mark or blemish on your fish. Take stock of the situation, consult reference sources, ask advice and do not rush into inappropriate or unnecessary treatment.

478 Quarantine should be mandatory

Due to the amount of rockwork in a reef system, it is often virtually impossible to catch the fish established in it. An infected fish can very easily pass on disease organisms to the other fish, and given the difficulty of treating fish disease in a reef setting, this can quite easily lead to a fish wipeout. This is why quarantining new fish is so vital.

Tip 477 *Get to know your fish so you will quickly recognize any changes in their normal behavior.*

A quarantine tank

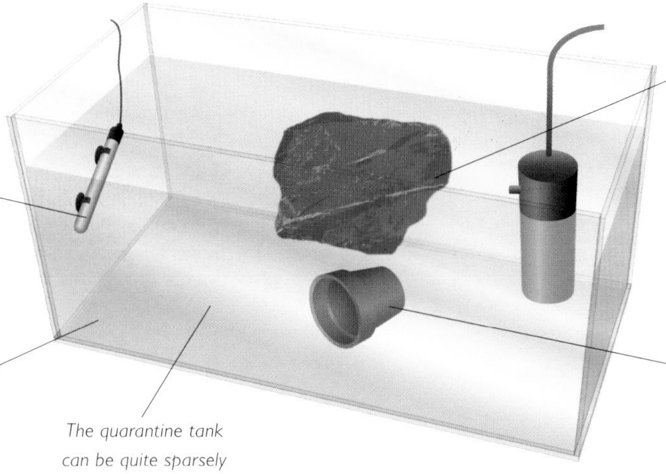

Use a heaterstat to maintain water temperature. Install a guard to protect fish that might rest against the heater element.

Slate is an inert material and will not affect the water quality.

Use a clean tank with no substrate.

The quarantine tank can be quite sparsely furnished.

A clay flowerpot provides a useful hiding place.

479 Setting up a simple quarantine tank

A suitable aquarium can double as a small quarantine/treatment tank. The tank can generally be quite small, perhaps 18–24 gallons (68–90 L) or so. A fully furnished tank, with calcareous rockwork or gravel, is not conducive to treatment, and neither is a completely bare tank. The ideal setup is somewhere between these two extremes, namely a sparse tank with some rockwork to provide cover and a filtration system suited to treatment, if and when required.

480 Calcareous materials can affect medications

Dead coral used for tank decoration, coral sand and gravel, and other calcareous materials have the ability to adsorb chemicals from the tank water and can diminish the effectiveness of a separate treatment tank. Use alternatives such as slate. Filtration could be provided by an internal power filter packed with suitable non-calcareous material, giving mechanical and biological filtration.

481 Try a short freshwater bath for new fish

Freshwater baths can be used as a prophylactic treatment for new fish additions. Any parasites will rupture under the osmotic shock of freshwater long before the fish experiences any harmful effects. Give the fish a two- to three-minute freshwater bath before adding it to the main tank. To lessen the shock to the fish, the freshwater must be at the same temperature and pH as your tank water. Methylene blue can also be added to enhance the effect of the bath. Formalin baths are nowadays generally regarded as too toxic.

482 Garlic wards off protozoan parasitic infections

Soaking the food of newly introduced fish in any of the commercially available garlic preparations can reduce the likelihood of an outbreak of a protozoan parasitic infection. Such diseases are commonly encountered in the aquarium as a result of the stress the fish endure when they are caught, transported and introduced into their new home.

483 Rid corals of pest organisms using a freshwater dip

Freshwater dips, in water of the same temperature and pH as in your reef tank, can be used to rid corals of predatory nudibranchs and planarian flatworms *(Convolutriloba retrogemma)*. Submerge the coral with its rock into the dip and gently agitate for two to three minutes to help dislodge any parasites. In the case of nudibranchs, you may need to repeat the process at a later date if there are any egg masses present, as the dip will have no effect on the eggs. When the eggs hatch, the newly emerged nudibranchs will be affected by any freshwater dips.

484 Quarantine invertebrates as well as your fish

Quarantine is not just for fish. By observing newly acquired corals for 7–14 days, you have the chance of avoiding the introduction of potential pest organisms, such as *Aiptasia*, mantis shrimp, planarian flatworms, predatory snails, etc., into your display tank.

485 Alter the water parameters in a treatment tank

If treating for parasites in a separate quarantine tank, raise the tank temperature to 80–86°F (27–30°C), as this will speed up the life cycle of some parasites and consequently shorten the length of quarantine required. Parasite treatment can also be enhanced by low specific gravities. Parasites do not survive well at low specific gravities of 1.016–1.018. These are much lower than the levels of 1.022–1.024 that are usually maintained in fish-only tanks. If you decide to pursue these options, be sure to acclimatize your fish slowly as it returns from quarantine to the display aquarium.

486 Treating fish with a copper medication

If you need to treat a fish with copper in a quarantine tank, the usual treatment time is 14 days. If a second fish is added during this period, reset your clock to day one and continue treating *both* fish for the full 14 days.

Tip 484 *Quarantining reduces the risk of introducing planarian flatworms* (Convolutriloba retrogemma) *to the display aquarium.*

Tip 488 Cryptocaryon irritans *is a common parasitic infestation.*

487 Recognizing signs of disease in saltwater fish

If a fish persistently scratches or "flashes" at rockwork or on the base of the aquarium, it is a sign of irritation, probably parasitic. More obvious signs of disease are ragged fins, inflamed gills and ulceration or spots on the fins, body or gills. If the fish does succumb to disease, the problem will have to be reliably identified and treated appropriately. Be methodical about treating disease. Make sure your original diagnosis is correct before beginning treatment. For example, a fish gasping at the surface may not be infected with gill flukes but simply reacting to a low level of oxygen in the water.

488 Parasite infections spread rapidly in the aquarium

Keep in mind the speed at which an aquarium can be overwhelmed by disease. In their natural habitat many fish will have a few parasites, as these are held in check by the vast dilution of the oceans, fish movement and the attentions of cleanerfish. However, in the close confines of an aquarium, a parasite can complete its life cycle quickly and multiply at a tremendous rate, thus overwhelming the original carrier fish and passing quickly to its tankmates.

489 Note the actual volume of water in the aquarium

When filling the new aquarium for the first time, make a note of the actual volume of water used. Because of the displacement caused by aquarium decoration, plus the needs of the filtration system, this figure will not be the same as the theoretical capacity of the tank. You will need this information in order to dose the aquarium accurately in times of disease. Bear in mind that every time you add or remove a rock or coral you will need to take into account the change in the water volume.

490 A freshwater bath can be a useful disease treatment

A good basic therapeutic treatment for saltwater fish diseases is to give the infected fish a brief bath in freshwater. Dip length will be dictated by fish species (some fish are less tolerant than others) and by the disease to be treated. (For example, the parasitic infection caused by *Brooklynella* may need 15 minutes.) Keep a watchful eye on a fish during such treatment—remove it and place it in saltwater should it show signs of distress. These include thrashing around or the fish spitting water.

491 A mixed community tank is more difficult to medicate

Many hobbyists want to keep a mixed community of fish and invertebrates in the same aquarium. However, there can be problems, the most obvious of which is the treatment of disease. The most effective treatments for the most common saltwater fish diseases contain copper and are deadly to invertebrates. If disease is brought into the tank, you have no choice other than to treat the show tank, and thereby sacrifice the invertebrates, or to remove all the fish to a treatment tank.

492 Some remedies are safe with invertebrates

There are a few commercially available disease treatments that purport to be safe for use in an aquarium containing invertebrates. Some do indeed appear to be safe and give good results in certain cases, but they tend not to be as effective in general as more aggressive, invertebrate-toxic treatments. Others may have a deleterious effect on certain species of invertebrates. Read the packaging and all instructions to ensure that any medication is safe with your selection of invertebrates.

493 Follow directions, especially with regard to filtration

When using medication, read the instructions! Some medications may not be suitable for certain fish species. Remove activated carbon media from the filtration system during treatment, as it will remove the medication before a cure is achieved. Many medications also require you to suspend protein skimming for the duration of the treatment.

494 Do not add medications one after the other

If a treatment fails to work, never add another medication right after. First, some combinations of remedies may be toxic; and second, if the treatment succeeds, how will you know if it was the second remedy or a belated cure from the first one? If the treatment fails to work, be sure to return the water conditions in the treatment tank to normal before trying another remedy. After successful treatment, always restore the water conditions in the treatment tank to those of the main aquarium before returning the patient.

495 Sterilize the treatment tank and equipment after use

Clean and sterilize the treatment tank and all associated equipment after each and every use. To sterilize, either use commercially available aquarium products or those available for treating baby bottles, etc. Rinse well with freshwater afterward, then leave everything dry until required again.

Tip 492 *Some proprietary treatments can be used in a tank containing invertebrates. Always read the instructions.*

496 Fish are not equipped to deal with alien parasites

Mixing fish from different locations (e.g., Atlantic with Pacific species) can lead to problems with parasitic infections. The fish will have poorly developed defense mechanisms against the alien parasites, as they will not have encountered them before.

Tip 496 *Masked butterflyfish* (Chaetodon semilarvatus) *from the Indian Ocean.*

497 UV sterilization can deal with a number of diseases

UV sterilization is a very useful, noninvasive way of dealing with a number of diseases, both in the fish-only and the reef aquarium. Although anyone aspiring to be a better aquarist will quarantine any organism before introducing it into the display tank, this does not completely preclude the presence of disease organisms.

498 Continuous UV sterilization in a fish-only system

Fish-only systems may do better with the continuous use of UV sterilization, taking into account the higher stocking levels and resulting increased risk of stress due to territorial imperatives and aggression. Stress is probably the major cause of immune deficiency; any fish in this state will be wide open to any disease organisms present.

499 UV sterilization in the reef aquarium

A sensibly stocked reef system should be at lower risk of stress-induced disease, but given the difficulty of treating disease in a reef environment, incorporating a UV system can be seen as a form of insurance. It is probably not necessary to run UV sterilization nonstop; rather, turn on the system for a period of two to three weeks when adding new fish. Run the UV when you detect the presence of disease and continue to do so, again for a period of two to three weeks, after the last symptoms of disease have disappeared.

500 Humane euthanasia may become unavoidable

If all else fails, you may have no option but to dispose of a fish humanely. The best option is to obtain the anesthetic MS222 from a veterinarian. Follow the instructions closely and leave the fish in the solution for a few hours.

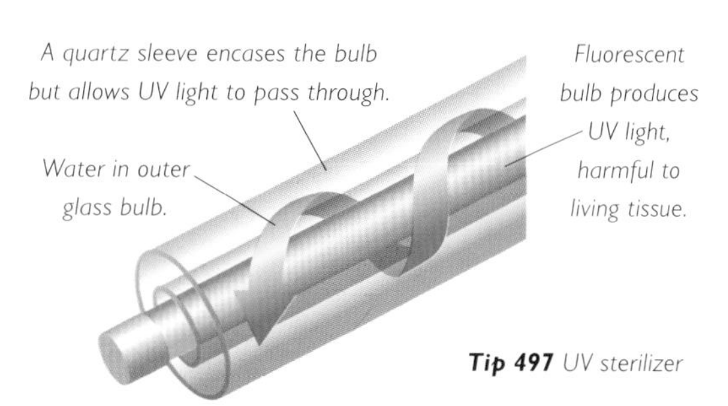

Tip 497 UV sterilizer

Credits

Unless otherwise stated, photographs have been taken by Geoff Rogers © Interpet Publishing.

The publishers would like to thank the following photographers for providing images, credited here by page number and position: (B) Bottom, (T) Top, (C) Center, (BL) Bottom left, etc.

Aqua Medic: 20(T)

Aqua Press (M-P & C Piednoir): 17(B)

Bioquatic Photo – A. J. Nilsen, NO-4432 Hidrasund, Norway (email: bioquatic@biophoto.net. Website: www.biophoto.net): 5, 6, 7(BL), 15(TR), 16(R), 21, 26, 27, 33, 35(TR), 39, 44(R), 46(BL), 49, 51(TR), 52, 53(TR,CL,BR), 54(TR), 55(TL), 57, 59(TR), 60(TR,CL), 61, 64(TR,BL), 65, 66(BR), 72, 74(CL,BR), 77(BR), 78(BL,CR), 79(TL,BR), 80(CL,CR), 81(TL,BR), 82(TL,BR), 83, 84(TR), 85, 86(CL,BR), 87, 91(TR), 92(BL,CR), 94(TR), 96, 104, 105(CL,TR), 106(CL), 108, 115(CR), 120, 121(CL,BL,BR), 124

D-D Aquarium Solutions: 20(BL), 32(L), 103(TR)

GHL Products: 16(CL), 50(CL)

Tim Hayes: 35(CL), 112(Inset), 114 (BL,BC,BR), 115(BL), 118, 119(CL,CR)

Tristan Lougher: 63(BR)

Photomax (Max Gibbs): 125

David Stephens: Copyright page, 38, 75, 110(T), 113, 127(CL)

Computer graphics by Phil Holmes and Stuart Watkinson © Interpet Publishing

The Authors

Dave Garratt has been keeping saltwater fish for more than 25 years. He is a regular contributor to hobbyist magazines such as *Tropical World* and *Tropical Fish*, and specializes in articles aimed at beginners.

Tim Hayes writes extensively on reefkeeping for a variety of aquatic magazines. He is a regular contributor to *Practical Fishkeeping*, where he has a monthly column and is a key member of an expert advisory team.

Tristan Lougher is a qualified zoologist with a passion for saltwater aquariums. His role in the retail side of the ornamental fish industry enables him to give and receive vital fishkeeping advice. He is a regular contributor to *Marine World* magazine.

Dick Mill's abiding interest in aquarium fish has lasted at least 40 years. He writes for monthly hobbyist magazines, lectures frequently and has written many fishkeeping books. He was invited to judge at Singapore's Aquarama in 2005.

Acknowledgements

The publishers would like to thank the following for their help: Arcadia, Croydon, Surrey; Nick Crabtree, Coral Culture; Liz Donlan, *Marine World* magazine; Tuan Pham; Seapet Centre, Martlesham Heath, Suffolk; Sevenoaks Tropical Marine, Sevenoaks, Kent; Swallow Aquatics, Colchester, Essex; Swallow Aquatics, Rayleigh, Essex; Swallow Aquatics, East Harling, Norfolk.

Publisher's note

The information and recommendations in this book are given without guarantee on the part of the author and publishers, who disclaim any liability with the use of this material.